Voice and Environmental Communication

Palgrave Studies in Media and Environmental Communication

Series Editors: **Anders Hansen**, University of Leicester, UK and **Stephen Depoe**, University of Cincinnati, USA.

Advisory Board: **Stuart Allan**, Cardiff University, UK, **Alison Anderson**, Plymouth University, UK, **Anabela Carvalho**, Universidade do Monho, Portugal, **Robert Cox**, The University of North Carolina at Chapel Hill, USA, **Geoffrey Craig**, Universtity of Kent, UK, **Julie Doyle**, University of Brighton, UK, **Shiv Ganesh**, Massey University, New Zealand, **Libby Lester**, University of Tasmania, Australia, **Laura Lindenfeld**, University of Maine, USA, **Pieter Maeseele**, University of Antwerp, Belgium, **Chris Russill**, Carleton University, Canada and **Joe Smith**, The Open University, UK.

Global media and communication processes are central to how we know about and make sense of our environment and to the ways in which environmental concerns are generated, elaborated and contested. They are also core to the way information flows are managed and manipulated in the interest of political, social, cultural and economic power. While mediation and communication have been central to policy-making and to public and political concern with the environment since its emergence as an issue, it is particularly the most recent decades that have seen a maturing and embedding of what has broadly become known as environmental communication.

This series builds on these developments by examining the key roles of media and communication processes in relation to global as well as national/local environmental issues, crises and disasters. Characteristic of the cross-disciplinary nature of environmental communication, the series showcases a broad range of theories, methods and perspectives for the study of media and communication processes regarding the environment. Common to these is the endeavour to describe, analyse, understand and explain the centrality of media and communication processes to public and political action on the environment.

Titles include:

Alison G. Anderson
MEDIA, ENVIRONMENT AND THE NETWORK SOCIETY

Jennifer Peeples and Stephen Depoe (*editors*)
VOICE AND ENVIRONMENTAL COMMUNICATION

Palgrave Studies in Media and Environmental Communication
Series Standing Order ISBN 978–1–137–38433–1 (hardback)
978–1–137–38434–8 (paperback)
(*outside North America only*)

You can receive future titles in this series as they are published by placing a standing order. Please contact your bookseller or, in case of difficulty, write to us at the address below with your name and address, the title of the series and the ISBN quoted above.

Customer Services Department, Macmillan Distribution Ltd, Houndmills, Basingstoke, Hampshire RG21 6XS, England

Voice and Environmental Communication

Edited by

Jennifer Peeples
Utah State University, USA

Stephen Depoe
University of Cincinnati, USA

Introduction, selection and editorial matter © Jennifer Peeples and Stephen Depoe 2014
Individual chapters © Respective authors 2014

All rights reserved. No reproduction, copy or transmission of this publication may be made without written permission.

No portion of this publication may be reproduced, copied or transmitted save with written permission or in accordance with the provisions of the Copyright, Designs and Patents Act 1988, or under the terms of any licence permitting limited copying issued by the Copyright Licensing Agency, Saffron House, 6–10 Kirby Street, London EC1N 8TS.

Any person who does any unauthorized act in relation to this publication may be liable to criminal prosecution and civil claims for damages.

The authors have asserted their rights to be identified as the authors of this work in accordance with the Copyright, Designs and Patents Act 1988.

First published 2014 by
PALGRAVE MACMILLAN

Palgrave Macmillan in the UK is an imprint of Macmillan Publishers Limited, registered in England, company number 785998, of Houndmills, Basingstoke, Hampshire RG21 6XS.

Palgrave Macmillan in the US is a division of St Martin's Press LLC, 175 Fifth Avenue, New York, NY 10010.

Palgrave Macmillan is the global academic imprint of the above companies and has companies and representatives throughout the world.

Palgrave® and Macmillan® are registered trademarks in the United States, the United Kingdom, Europe and other countries.

ISBN 978–1–137–43373–2

This book is printed on paper suitable for recycling and made from fully managed and sustained forest sources. Logging, pulping and manufacturing processes are expected to conform to the environmental regulations of the country of origin.

A catalogue record for this book is available from the British Library.

Library of Congress Cataloging-in-Publication Data
Voice and environmental communication / edited by Stephen Depoe,
   University of Cincinnati, USA; Jennifer Peeples, Utah State University, USA.
      pages cm — (Palgrave studies in media and environmental communication)
   Summary: "Voice and Environmental Communication explores how people give voice to, and listen to the voices of, the environment. As anxieties around degrading environments increase, so too do the number and volume of voices vying for the opportunity to express their experiences, beliefs, anxieties, knowledge and proposals for meaningful change. Nature itself speaks through, and perhaps to, individuals who advocate on behalf of the environment. This collection includes nine original essays organized into three sections: Voice and Environmental Advocacy, Voice and Consumption, and Listening to Non-human Voices. Four notable scholars reflect on these chapters, and provide both an audience to the scholars as well as a forum for extending their own understanding of voice and the environment. This foundational book introduces the relationship between these two fundamental aspects of human existence and extends our knowledge of the role of voice in the study of environmental! communication."— Provided by publisher.
   ISBN 978–1–137–43373–2 (hardback)
   1. Communication in the environmental sciences.   2. Mass media and the environment.   I. Depoe, Stephen P., 1959– editor.   II. Peeples, Jennifer Ann, editor.
   GE25.V65 2014
   363.7001′4—dc23   2014019735

Contents

Notes on Contributors   vii

Introduction: Voice and the Environment—Critical
Perspectives   1
Jennifer Peeples and Stephen Depoe

Section I   Voice and Environmental Advocacy

1   Corporate Ventriloquism: Corporate Advocacy, the Coal
Industry, and the Appropriation of Voice   21
*Peter K. Bsumek, Jen Schneider, Steve Schwarze,
and Jennifer Peeples*

2   Defending the Fort: Michael Crichton, Pulp Fiction, and
Green Conspiracy   44
Patrick Belanger

3   Invoking the Ecological Indian: Rhetoric, Culture,
and the Environment   66
Casey R. Schmitt

4   Sustainable Advocacy: Voice for and before an
Intergenerational Audience   88
Jessica M. Prody and Brandon Inabinet

5   Response Essay: The (Im)possibility of Voice in
Environmental Advocacy   110
Danielle Endres

Section II   Voice and Consumption

6   Voices of Organic Consumption: Understanding Organic
Consumption as Political Action   127
Leah Sprain

7   *Vote with Your Fork*: The Performance of Environmental
Voice at the Farmers' Market   148
Benjamin Garner

8 Response Essay: Thinking through Issues of Voice
   and Consumption   170
   Laura Lindenfeld

Section III Listening to Nonhuman Voices

9 The Language That All Things Speak: Thoreau and the
   Voice of Nature   183
   William Homestead

10 The Ethics of Listening in the Wilderness Writings
   of Sigurd F. Olson   205
   David A. Tschida

11 Listening to the Natural World: Ecopsychology of
   Listening from a Hawai'ian Spiritual Perspective   228
   Yukari Kunisue

12 Response Essay: Environmental Voices Including Dialogue
   with Nature, within and beyond Language   241
   Donal Carbaugh

13 Coda: Food, Future, Zombies   257
   Eric King Watts

Index   264

Contributors

Patrick Belanger is Assistant Professor of Communication and Transformative Conflict Resolution at California State University, Monterey Bay.

Pete Bsumek is Associate Professor in the School of Communication Studies at James Madison University.

Donal Carbaugh is Professor of Communication and Director of the Graduate Program in Communication at the University of Massachusetts Amherst.

Stephen Depoe is Professor and Head of the Department of Communication, University of Cincinnati.

Danielle Endres is Associate Professor of Communication and faculty in the Environmental Humanities Masters Program at the University of Utah.

Benjamin Garner is Assistant Professor, Department of Marketing, University of North Georgia.

William Homestead is Assistant Professor in the Department of Communication Studies at New England College.

Brandon Inabinet is an Assistant Professor of Communication Studies at Furman University, specializing in the ethics of rhetoric and argumentation and whose work has appeared in *Advances in the History of Rhetoric, Southern Communication Journal, Rhetoric & Public Affairs,* and *Rhetoric Society Quarterly* and coordinator of David E. Shi Center for Sustainability Affiliate Faculty.

Yukari Kunisue is Acting Director of Student Services, Hawaii Tokai International College.

Laura Lindenfeld is Associate Professor in Communication and Journalism at the Margaret Chase Smith Policy Center at the University of Maine.

Jennifer Peeples is Associate Professor of Communication Studies at Utah State University.

Jessica M. Prody is Assistant Professor of Performance and Communication Arts at St. Lawrence University.

Jen Schneider is Associate Professor in Liberal Arts and International Studies at the Colorado School of Mines in Golden, Colorado.

Casey R. Schmitt is a PhD candidate and Lecturer in Rhetoric, Politics, and Culture at the Communication Arts department of the University of Wisconsin-Madison.

Steve Schwarze is Chair and Associate Professor of Communication Studies at the University of Montana at Missoula.

Leah Sprain is Assistant Professor at the University of Colorado Boulder.

David A. Tschida is Assistant Professor of Communication at the University of Wisconsin-Eau Claire, an affiliate faculty member in the UWEC Watershed Institute for Collaborative Environmental Studies.

Eric King Watts is Associate Professor in the Department of Communication Studies at the University of North Carolina at Chapel Hill.

Introduction: Voice and the Environment—Critical Perspectives

Jennifer Peeples and Stephen Depoe

> Society speaks and all men listen, mountains speak and wise men listen
>
> —John Muir

Clean water, air, and soil. Wild and open spaces. Uncontaminated foods. Healthy bodies and communities. These are some of the scarce resources that come to mind when thinking about environmental issues. And yet there is another limited resource, one that is intricately tied to the environment and yet often not recognized as such: voice. While there is often a cacophony of people talking, what is missing is the acknowledged voice, the one that is given an audience, allowed to be impactful and transformative in its assertions—the one that is heard. As Couldry (2010) warns, voice is in crisis. We daily witness the devastation aided by the loudly expressed agendas of a small minority of people who are able to dictate the environmental outcomes for the majority. As we maintain in this book, as voice goes, so goes the environment.

In the following chapters we explore the ways people give voice to, and listen to the voices of, the environment. Voice is not simply analogous to speaking; it is the "enunciation and the acknowledgement of the obligations and anxieties of living in community with others" (Watts, 2001, p. 180). And in the case of environmental concerns, whose breadth and magnitude affect every living thing on the planet, the circle of "community" is quite large. In the first chapter of his book *Environmental Communication and the Public Sphere*, Cox (2013) lists the individuals and organizations involved in environmental conflict: citizens and community groups, environmental groups, scientists, corporations and business lobbyists, anti-environmentalist groups, media and environmental journalism, and public officials and regulators (pp. 26–32). Each has a voice and each attempts to find a receptive audience. As our anxieties around our changing environments increase, so too do the number and volume of the environmental voices vying for an opportunity to express their experiences, their beliefs, their fears, their knowledge and their proposals for meaningful change. Nature itself, it may be argued, is speaking through, and perhaps to, individuals who advocate on behalf of various environmental causes.

Our text delves into the multifaceted nature of voice, recognizing that voice is power—it can be given and taken away. It has the capacity to create presence, it is used as a means to oppress or resist, as a response to alienation, and it is the sound of becoming (Watts, 2001). Like the environment, voice is socially grounded and conditioned by its cultural, political, economic, and historical contexts (Brady, 2011, p. 203). Finally, as environmental decisions are always contested and often contentious, voice is the currency of environmental struggle.

Within the communication discipline, there is an implicit understanding of the importance of voice for environment issues. Texts such as Shaiko's (1999) *Voices and Echoes for the Environment: Public Interest Representation in the 1990s and Beyond*; Muir and Veenendall's (1996) *Earthtalk: Communication Empowerment for Environmental Action*, and Killingsworth and Palmer's (1992) *Ecospeak: Rhetoric and Environmental Politics in America* all point to voice's significance. Some specifically address voice, such as Senecah's "Trinity of Voice" essay (2004), while others touch on various aspects of voice found under the broad umbrella of environmental communication. As of now, no volume has taken up the concept of voice as its primary focus. In response, this book explores the multidimensionality of voice in order to understand its functioning given the particular constraints found within environmental issues. Our book is in no way intended to be the final word on this complex subject, but is an effort to illuminate this vital aspect of environmental communication.

As introduction, we lay out five aspects of voice integral to understanding its impact on environmental concerns and that provide a conceptual framework that underlies the arguments that follow. We present examples of environmental scholars who have directly or indirectly incorporated aspects of voice into their scholarship. We end with a preview of the chapters and reflections that expand upon these essential elements of voice.

This book is not only an explanation of voice, but also an enactment of it. In addition to the two editors, we have nine authors writing on specific and diverse enactments of voice and the environment. These authors are presented in conversation with four noted scholars who reflect on what they have heard in these chapters, providing both an audience for these authors and a means of extending their own thoughts and arguments concerning environmental communication and voice.

Voice and identity

At its most basic level, voice is a physiological process, a mechanism for expressing one's thoughts through sound and action. It is also the "instrument, the vehicle, the medium" for constructing meaning for ourselves and others (Dolar, 2006, p. 4). As Appelbaum (1990) contends, "voice, sound and meaning are so commingled as to make a natural unity" (p. 4). For many, the process of voicing an idea is what allows for understanding, as one is forced to choose one symbol over another in order to assemble a particular perspective.

Voice is also commonly associated with the expression of an opinion or the articulation of a "distinctive perspective on the world" (Couldry, 2010, p. 1). In his analysis, *Why Voice Matters*, Couldry contends that there is one primary purpose for using one's voice: it is the "process of giving an account of one's life and its conditions" (2010, p. 7). We would add that through the practice of giving an account, the speaker is also constructing his or her identity, place, and life experiences. The expression of and the constitution of a life story are intertwined and inseparable throughout the process of giving voice. Couldry concludes with the warning that to deny a person's potential for voice "is to deny a basic dimension of human life" (2010, p. 7).

In environmental movements, individuals with distinct perspectives on the natural world, and the human impact upon it, have shaped how people understand and interact with their environments. But these perspectives might have remained unknown had these impressive thinkers not had equally impressive and distinctive voices for change: the lyrical prose of John Muir, the storytelling of Aldo Leopold (Meine, 1999), the scientific narratives of Rachel Carson, and the sociological articulations of Robert Bullard, among others. Each voice captivated audiences, garnering attention to their influential perspectives. Because of the deep environmental impact of these and other key individuals, communication scholars have investigated how their discourse was able to influence audiences when that of so many other like-minded and equally knowledgeable people was not. Examples of communication scholarship focusing on the voices of environmental advocates include Oravec's work on John Muir (1981), Ullman's examination of

Aldo Leopold (1996), Waddell's edited volume on Rachel Carson (2000), Hope's comparison of the autobiographical voices of Lois Gibbs and Sandra Steingraber (2004), Rosteck and Frentz's essay on Al Gore (2009), Singer's analysis of Thomas Friedman (2010), and Gorsevski's study of the emplaced rhetoric of Kenyan activist Wangari Maathai (2012).

Voice as textual, or intertextual

Voices do not emanate merely from persons, but also from within and between texts; so says literary theorist Mikhail Bakhtin (1981, 1984, 1986). According to Phillips, Carvalho, and Doyle (2012), Bakhtin "understands a voice not just as the medium for speech or the uttered speech of an individual, embodied person, but as a discourse, ideology, perspective or theme that transcends the individual" (p. 7). From this perspective, multiple voices may be present or at work within a single text or discourse (polyphony), and multiple and even conflictual meanings may be discerned based on attention paid to the voice or voices inhering within or among texts (heteroglossia). Extending earlier work on voice to the realm of cyberspace, Mitra and Watts (2002) note that voice "operates as a *sign* of a set of cultural meanings at work in a social body," and "need not be bound to any geopolitical space or social location" (p. 483). Voice in a digital age can be viewed as "synthetic," as a "dialogic event" or "happening" in which the production, dissemination, and reception or "hearing" of voices should be viewed as both "intertextual" and "mediated" (Mitra & Watts, p. 483). Bakhtin's approach allows our understanding of voice to broaden from the ontic (originating in a person) to the symbolic (existing as meaning and meaning-making).

A number of environmental communication scholars have adapted and extended Bakhtin's understanding of voice as a perspective or set of suggested meanings that operates within or among texts. In their landmark book *The Language of Environment: A New Rhetoric*, Myerson and Rydin (1996) coined the term "environet" to describe as an interactive flow of texts and voices unfolding across time and place following a discernible range of issues and events (such as endangered species or climate change). Employing the metaphor of a carnival to depict a dynamic system of changing meanings and connections across a society or community over time, a system comprised of texts and voices, Myerson and Rydin brilliantly foreshadow the ways in which contemporary communication networks produce and circulate words, sounds, images, and voices expressing positions on environmental (and other) issues. Complementing the macro- or system-level view of Myerson and Rydin, Marafiote and Plec (2006) examined the presence of diverse viewpoints or voices within discourses of individual people who expressed their views about the natural world through survey responses. In the data, Marafiote and Plec identified not only multiple voices, but the emergence of new combinations or hybrids of established ideological positions (both anthropocentric and ecocentric) on the environment, and concluded that the presence of both organic (unconscious) and strategic hybridity of voices may help to account for incongruities and conflicts in environmental discourses and debates. Moving from the level of the individual to the group or community, Hamilton (2007) and Hamilton and Wills-Toker (2006) have employed Bakhtin to examine how interactions evolve among participants in various public participation formats pertaining to environmental policymaking, including public hearings and community advisory boards. These authors examined how particular points of view (voices) are articulated and circulated among stakeholders and government officials. This line of inquiry has been extended to the realm of risk and science communication in a recently published work entitled, aptly enough, *Citizen Voices: Performing Public Participation in Science and Environment Communication* (Phillips, Carvalho, & Doyle, 2012).

Other scholars have examined the textual construction of voices by those who are interested more in material acquisition and profit than in environmental protection. Peeples (2005) has examined how pro-business (coined as "wise use") advocacy groups attempt to imitate the voice of environmental advocates in order to thwart or blunt the impact of those opposing views. Plec and Pettenger (2012) have analyzed the ways in which ExxonMobil has projected a benign voice (also referred to as "greenwashing") in their advocacy of various energy solutions that are consistent with their own corporate bottom line.

Voice and social organizing

As with other social, political, and/or cultural controversies, not all points of view in environmental controversies are deemed permissible or significant enough for inclusion. Dissenting voices are separated from decision-makers in "protest zones," demarcated by chain link fences or stricken from public records. For Couldry, the most obvious reason a voice is excluded from a discussion is a practical one: an entity lacks a language with which to articulate its situation. This is especially problematic for those elements of an ecosystem that are not able to express their circumstances through a human symbol system, a point we will come back to later in this chapter.

The second reason is more complex in that it is structural and systemic. A person must have the necessary "status" if "one is to be recognized by others as having a voice" (Couldry, 2010, p. 7). And status requires an audience. Watts (2012) contends that "voice does not occupy the private body for very long; it seeks a hearing and often 'dies' before receiving one. A condition for voice, thus, is social" (p. 16). For the process to be complete, voice requires "both speaking *and listening*, that is, an act of attention that registers the uniqueness of another's narrative" (Couldry, 2010, p. 9). Voice is thus "actualized by public acknowledgment" (Watts, 2001, p. 186).

Couldry warns that voice is also undermined by systems "which take no account of voice" (2010, p. 10). He points to a current crisis where voices "are increasingly unsustainable; voice is persistently offered, but in important respects denied or rendered illusionary" (2010, p. 1). Above all, he argues, "voice is undermined when societies become organized on the basis that individual, collective and distributed voice need not be taken into account, because a higher value or rationality trumps them" (2010, p. 10).

In response to entities that systematically exclude oppositional voices, people with environmental concerns are often motivated to find audiences for their voices outside the formalized strictures of the government processes and other sanctioned acts of public address. It is voice that allows for the formation of organizations and groups, as it functions at the intersection of individuality, subjectivity, and connection with others (Dolar, 2006, p. 4). The social aspect of voice allows for the construction of commonality and community; upon hearing the stories of others, people find similarity with their own lives (Hauser, 1999).

Individuals form social movements, direct actions, and nonprofit organizations to counter the silencing they feel as they attempt to espouse the dire changes they have witnessed in the environment (Stewart, Smith, & Denton, 2001; Stillion Southard, 2007). Organized political activity related to environmental issues has taken many forms, from referendum campaigns to protests to "eco-tage" in the name of halting harmful industrial projects (Lange, 1990; Shabecoff, 2003). Groups have also taken to organizing around consumer choices, instituting boycotts and "buy-cotts" in an effort to put economic pressure on companies that produce and distribute products whose manufacture and consumption adversely impact the quality of the biosphere, including human and nonhuman indicators (Micheletti, 2003; Pezzullo, 2011).

In recent years, some of the most influential grassroots organizations in environmental disputes have brought together issues of race, socio-economic status, and environmental concerns under the unifying umbrella of environmental justice (Sandler & Pezzullo, 2007). From "hysterical housewives" to communities of color vowing to "speak for themselves" (Alston, 1990; Zeff, Love, & Stults, 1989), the voices of the margin have been able to change how the public thinks about issues of race and the environment. In their analysis of the use of the feminine style and material militancy in the discourse of the environmental justice movement, Peeples and DeLuca (2006) explore how individuals come together into powerful coalitions capable of altering the grounds upon which environmental decisions are made. Specifically, women involved in environmental justice often describe how they formed organizations after hearing other women voice concerns about their children's health, which resonated with their own experiences. In finding that they were not alone, they were empowered to take action to clean up the toxins in their neighborhoods. One of the strategies these environmental justice advocates use is to unite under the banner of "motherhood" (Peeples & DeLuca, 2006). They argue that no one (no politician, no epidemiologist) has greater knowledge of their children or their communities than they do, establishing their status as experts. As scientific methods often prove inconclusive in issues of toxins, especially in the small sample sizes of some of these communities, the women use their collective knowledge of their children's health and their own body epistemology, along with their community knowledge of dump locations and neighborhood disease clusters, to question the scientific findings that disagree with their own experiences (Peeples & DeLuca, 2006). The successes of environmental justice advocates in raising awareness of these issues comes in no small measure from individuals voicing their lived experiences and using the similarities of those experiences to form organizations able to question the political and scientific authority.

Voice and political process

Voice has a particular significance under democratic governance. Huspek and Kendall define democracy as "a field of discursive struggle defined by political participants competing to get their words and meanings accepted by others in an effort to secure limited material and symbolic resources" (1991, p. 1). It is the variety of voices, stakeholders with competing interests, that (in theory) leads to the best possible outcomes. "Inextricably bound up with this view is the belief that both democracy and freedom may be diminished, even imperiled, when citizens withhold their voices from the formal political arena" (Huspek & Kendall, 1991, p. 1). Without all the interested voices, debate is limited or skewed, leaders may go "unchecked," and the impact of an individual's expression is reduced, as the rhetorical situation may shift away from that person's concerns. "Withholding one's political voice, therefore, amounts to a forfeiture of self-determination," a forfeiture of power (Huspek & Kendall, 1991, p. 1). In addition to individual agency, formal policies, structures, and common practices allow for or deny voices' ability to influence governmental processes.

In the United States, voice plays a unique and pivotal role in environmental decision-making. The National Environmental Policy Act (NEPA) of 1970 requires a public review of environmental impact statements for projects that might cause environmental harm. A number of state and federal laws require opportunities be made for the public to participate in the process. This can be in the form of an open comment period or, equally common, a public hearing (Hendry, 2010, pp. 219–236).

While NEPA and related policies present a somewhat unprecedented opportunity for the public voice in environmental decision-making, significant limitations often arise in implementation (Fiorino, 1990). Officials and administrators can view hearings as a waste of time, an antagonistic process, and perceive the public as uninformed or hysterical. The citizen voices can be labeled as "indecorous," a term Cox (2013) uses to describe the "symbolic framing by some public officials of the voices of members of the public as inappropriate to the norm for speaking in regulatory forums and for the level of knowledge demanded by health and government agencies" (p. 255). Indecorous voices can be officially dismissed or informally ignored. On the other side, the citizen participants (and also some officials) often view the comment period as a charade, the façade of open and inclusive decision-making, with an outcome dictated far in advance of public involvement. Escalation of protest, they argue, is the only available means to make themselves heard, and this provides further evidence of their "irrationality" for officials.

Senecah (2004) proposes a "practical theory" of the Trinity of Voice (TOV) to evaluate a participatory process's ability to provide agency for citizen voices in decision-making. She maintains that for voice to have impact, or a hearing as stated by Watts (2012), within a given participation format, three conditions must be present: access, standing, and influence. Access is the opportunity to express one's choices and opinions. Standing is the "respect, the esteem and the consideration" of the stakeholders' perspectives (p. 24). Finally there is influence, which does not mean that the person's opinions or stated course of action is enacted, but that the person's ideas are given equal weight with others in the decision process. Many of the current methods of citizen participation allow access, but lack the provision of standing and influence. Senecah (2004) argues: "Stakeholders who are denied any of these components will find a means by which to claim them. Often these means are disruptive, destructive, and unproductive in achieving decision effectiveness and social effectiveness" (p. 23). Senecah's TOV has been applied in a variety of case studies of public participation in environmental decision-making (Walker, Senecah, & Daniels, 2006; Klassen, 2011).

Voices of nonhuman nature

Following Aldo Leopold's example (1949), authors in this text expand the communities we inhabit to include all living things and their ecosystems, thereby questioning nature's capacity for "voice" and the ability for people to "hear" nature's diverse ways of communicating. This approach raises a number of questions. What is nature's voice? Does it "speak"? If so, how? To whom? Can humans attend to the voice of nature? These questions stretch the theoretical and material dimensions of how we understand the relationship between voice and the environment. This extension of voice is crucial to understanding environmental issues, as many of the entities at the center of the conflicts do not speak, at least using a system of symbols that humans recognize as language (Lovelock, 1995; Hella & Dyke, 2006). The land, the air, the water, the mountains, the plains, the glaciers, the plants, the nonhuman animals, all have a stake in environmental outcomes that is equal to, if not more pressing than, that of their human cohabitants, but lack traditional voices to express their interests. Or does nature speak, if only we would listen (Abram, 1996; Capra, 1996; Haraway, 2008; Roszak, 1992)?

A number of communication scholars have written on the question of nature's voice. Rogers (1998) has proposed a "transhuman" model of communication, based on the argument that most Western rhetorical and communication theories that foreground humans' so-called unique powers of symbol-making and symbol-using are both simplistic and destructive. According to Rogers, the human move toward objectification of nature, ignoring its voice, seeing ourselves as stewards above and separate from it, erases any chance for dialogue with nature: "Ideologies of domination and manipulation, in effect, silence nature by dismissing the value of interaction based on nonhierarchical arrangements" (1998, p. 245). Rogers' model has been extended to examine how humans come to understand and talk about encounters with other species such as whales (Milstein & Krolokke, 2012).

Adopting a culture-centered approach, Carbaugh (1999) offers the view that Blackfeet people in North America have developed skills of discernment that allow them to "just listen" to the voice of nature in their particular places of habitation. Building on this perspective, Salvador and Clarke (2011) explore another indigenous way of attending to the voices of nature, of reconnecting humans to their nonhuman, nonsymbol-using others through the Nez Perce concept of the weyekin. The weyekin is the spiritual strength that people can actualize through physical and metaphysical interaction with the nonhuman in nature. Salvador and Clarke offer that "the power of the weyekin comes not from transcending (symbolizing nature) and the material world, but rather from a fully embodied dialogue with the nonhuman" (p. 248). The "listening" they prescribe requires an individual to allow nature to "resonate" within their bodies, as does music, as a means of moving from the symbolic to the corporeal. The second element of embodied listening requires mimicry, an imitative process established through repeated observations of the nonhuman other over time. Resonance and mimicry enable a person to listen beyond symbols, but requires "close attention, rigorous observation, and embodied presence" in nature (p. 251), thus allowing the listener to address the symbolic–material tension that so often separates humans from the natural, found within their bodies and in their surroundings.

Most recently, in the introduction to an edited volume on human–animal communication, Plec (2013) posits an "internatural communication" perspective that would include "the exchange of intentional energy between humans and other animals as well as communication and other forms of life" (p. 6). Plec hopes that her volume will encourage readers to rethink "our anthropocentric grip on the symbolic" by "becoming students of corporeal rhetorics of scent, sound, sight, touch, proximity, position, and so much more" (p. 7).

This volume hopes to contribute to the emerging conversation by focusing on how individuals, cultures, and communities listen (or fail to listen) to voices of nature in particular contexts. As a parting thought on this topic, cultural historian Thomas Berry offers the following insight:

The universe is composed of subjects to be communed with, not objects to be exploited. Everything has its own voice. That's why primordial peoples have a deep sense of relatedness to all natural phenomena. Thunder and lightning and stars and planets, flowers, birds, animals, trees—all these have voices, and they constitute a community of existence that's profoundly related.

(quoted in Jensen, 2002/2004, p. 36)

Berry goes on: "Our primordial spontaneities, which give us a delight in existence and enable us to interact creatively with natural phenomena, are being stifled. Somehow we have become autistic. We don't hear the voices" (2002/2004, p. 36).

In full agreement with Berry, a motivation for us in editing this book is the recognition that we do not hear the voices that surround us, whether they come from humans or from nature. This book explores what barriers are in place that preclude the hearing of voices, how voices are used in ways that we might not initially identify as articulating life experiences, and how we can use our voices and/or be an audience to the voices of others.

The chapters

Voice has an integral connection to communication studies. But, as noted above, environmental concerns, as opposed to other social, cultural or political issues, enact a particular set of constraints that affect voice. The environmental concerns affect everyone (though not equally), and often in essential ways: what a person eats or drinks or where one lives. Environmental problems, especially around toxins, directly and powerfully impact the poorest and most marginalized populations, making the question of voice and agency central to environmental concerns. Finally, many of the key players in environmental issues, including nature itself, do not have a commonly recognized "voice" with which to represent themselves and require human "translation" if they are to be influential in decision-making.

We took these fundamental environmental constraints and used them as an organizing structure for the sections of this book: Voice and Environmental Advocacy, Voice and Consumption, and Listening to Nonhuman Voices. The five themes of voice are articulated, examined, and intertwined throughout these three sections. Populating the sections are nine original essays written in response to our query as to how voice functions in environmental issues. Also, for each of the three sections we enlisted notable environmental communication scholars to reflect on how the concepts of voice raised in the chapters may function to influence our work in communication and our understanding of the environment.

In Part I, Voice and Environmental Advocacy, the authors consider aspects of voice and agency, voice and identity, and voice and opportunity that are integral to environmental activism. They address how voices shift and change in the course of advocacy, how voices are co-opted and used by others as well as asking how to advocate for people who are not yet present, but will be affected by current environmental decision-making. In "Corporate Ventriloquism: Corporate Advocacy, the Coal Industry, and the Appropriation of Voice," Pete Bsumek, Jen Schneider, Steve Schwarze, and Jennifer Peeples examine how the organizations supporting the coal industry, in an attempt to unify a range of people to "speak with one voice" about coal, engage in "corporate ventriloquism," a process by which corporations "throw" their voice through others to create the impression of broadly based support for coal. The authors unpack the implications of corporate ventriloquism for voice and environmental communication. In "Defending the Fort: Michael Crichton, Pulp Fiction, and Green Conspiracy," Patrick Belanger addresses the intersection of voice, narrative, and argument through questioning why the popular writer Michael Crichton was able to find a receptive audience for his explanation of climate change when others (with genuine expertise in environmental science) have not. In "Invoking the Ecological Indian: Rhetoric, Culture, and the Environment," Casey Schmitt uses the rhetoric of Winona LaDuke to reassess the discredited image of Native American as environmental prophet. He discusses the implications of the double bind faced by the Indian who is ecological and the simultaneous encouragement and discouragement of Native American voice that result from engagement with the Ecological Indian. Moving from current debate to future impacts, in "Sustainable Advocacy: Voice for and Before an Intergenerational Audience," Jessica Prody and Brandon Inabinet grapple with the question of how to protect, preserve, and better the world for future generations who as of yet do not have a voice in environmental decision-making. Drawing on Perleman's Universal Audience (UA), their essay seeks a rhetorical model that creates voice that is ethically constrained and compelled by sustainability, across all types of rhetorical situations. The section ends with a reflection piece from Danielle Endres explaining how the understanding of voice raised by the four authors may then be used to influence environmental decision-making and practice.

In Part II, Voice and Consumption, the authors examine how people's interaction with food, specifically their purchasing of organic goods and produce at farmers markets, "voices" a particular understanding of the human relationship with the environment. Voice and identity and voice and social and organizational processes play key roles in these discussions. In "Voices of Organic Consumption: Understanding Organic Consumption as Political Action," Leah Sprain considers how people conceptualize the meaning of organic consumption and how this consumption is rhetorically constituted in promotional materials. Her chapter demonstrates how organic consumption, paradoxically, is also a form of exit—or withdrawing one's voice—from political action. Benjamin Garner's "Vote with your Fork: The Performance of Environmental Concern at the Farmers' Market" suggests that for patrons, supporting the farmers' market and constituent farmers represents an expression of voice whereby customers display their environmental concerns not only by purchasing goods from farmers, but also by placing their physical bodies in a public place in a performative, embodied way. Our critical respondent for this section, Laura Lindenfeld, uses her extensive knowledge of food and communication to position the two voice chapters within the disciplinary discussions of consumption.

In Part III of the volume, Listening to Nonhuman Voices, the authors conceptualize different ways that people may be able to hear the voice of nature and critically investigate whether that voice can or should be translated into a human system of symbols. William Homestead, in "The Language that All Things Speak: Thoreau and the Voice of Nature," argues that Thoreau clears a path for narratives that integrate mind and nature, redefining human development within a spiritual and Earth context as well as a human capacity for listening to the voices of nature. In "The Ethics of Listening in the Wilderness Writings of Sigurd F. Olson," Tschida establishes Olson as an exemplar of how a common human-to-human communicative practice—dialogic listening—may serve as a means of ethically responsible human-to-nature relations when understood in a dialogic context. Finally, Kunisue explores the Hawaiian way of communication in relation to listening to nature in "Listening to the Natural World: Ecopsychology of Listening from a Hawai'ian Spiritual Perspective." She maintains that by encountering and applying Hawaiian concepts such as "*ha*" or tradition of "*Ho'oponopono*," scholars of listening will find a tool in communication to understand voice in nature and beyond modern human consciousness. Donal Carbaugh, drawing on his own extensive body of work on listening to the environment, provides a reflection essay for this final section of the book. Rhetoric scholar Eric Watts then closes the volume with a thought-provoking coda on the possibilities and challenges of finding voice in contemporary environmental advocacy put forth by humans, as well as from the Earth itself.

As we have constructed this volume as an introductory investigation into the articulation of voice and the environment, we can envision many ways for future scholars to extend and enrich the work found here. Following the path laid by existing voice research in the communication discipline, all of our authors have chosen humanist approaches for their analyses, with the majority using rhetorical methodologies and a few informing their work with performance, ethnographic, and ecopsychological approaches. As future environmental voice scholars extend Couldry's examination of modern media systems' influence on voice, and as examples broaden out from the United States, we see the opportunity for critical, cultural, and global perspectives to advance the discussions found within our volume. We understand our approach to research and subject matter as laying the foundation for further investigations into and conversations about voice and the environment.

Voice is a powerful tool, one that requires caution in both its expression and reception. Once a person responds to a voice's call, hears the voice of the other, a choice must be made as to whether to act on what the person now knows. Specifically for this project, as we hear the voices calling for change in the environment, or recognize the voice of the environment itself, we can no longer claim to be unaware of that voice's existence—that entity's story—as we make our decisions. As Watts astutely notes, "One does not necessarily embrace [the] intensity" of hearing another's voice (2012, p. 17).

And yet, if we are going to make changes, changes that positively impact all those within our environmental communities, we have to be able to listen to the voices of those who suffer under the current system of environmental decision-making, where economic gains are more important than human health and ecological sustainability. We have to listen to the voice of nature, which continues its unremitting distress signal that the natural systems are broken. And we have to listen for the silences where there once was a voice and there is no longer. And finally, we must add our voices to the cacophony of noise that surrounds so many of our important decisions. To refuse to do so is to forfeit our power and to deny a capacity, but more importantly, the responsibilities that come with being human.

References

Abram, D. (1996). *The Spell of the Sensuous: Perception and Language in a More-than-Human World*. New York: Pantheon Books.
Alston, D. (Ed.) (1990). *We Speak for Ourselves: Social Justice, Race, and Environment*. Washington, DC: Panos Institute.
Appelbaum, D. (1990). *Voice*. New York: State University of New York Press.
Bakhtin, M. (1981). *The Dialogic Imagination*. Tr. by C. Emerson. M. Holquist (Ed.). Austin and London: University of Texas Press.
Bakhtin, M. (1984). *Problems of Dostoevsky's Poetics*. Tr. by C. Emerson. Minneapolis: University of Minnesota Press.
Bakhtin, M. (1986). *Speech Genres and Other Late Essays*. Austin: University of Texas Press.
Brady, M. J. (2011). Mediating indigenous voice in the museum: Narratives of place, land, and environment in new exhibition practice. *Environmental Communication: A Journal of Nature and Culture, 5*, 202–220.
Capra, F. (1996). *The Web of Life: A New Scientific Understanding of Living Systems*. Garden City, NY: Anchor Books.
Carbaugh, D. (1999). "Just listen": "Listening" and landscape among the Blackfeet. *Western Journal of Communication, 63*, 250–270.
Couldry, N. (2010). *Why Voice Matters: Culture and Politics after Neoliberalism*. Los Angeles: Sage.
Cox, J. R. (2013). *Environmental Communication and the Public Sphere* (3rd ed.). Thousand Oaks: Sage Publications.
Dolar, M. (2006). *A Voice and Nothing More*. Cambridge, MA: MIT Press.
Fiorino, D. J. (1990). Citizen participation and environmental risk: A survey of institutional mechanisms. *Science, Technology, and Human Values, 15*, 226–243.
Gorsevski, E. W. (2012). Wangari Maathai's emplaced rhetoric: Greening global peacebuilding. *Environmental Communication: A Journal of Nature and Culture, 6*, 290–307.
Halla, Y., & Dyke, C. (Eds.) (2006). *How Nature Speaks: The Dynamics of the Human Ecological Condition*. Durham, NC: Duke University Press.
Hamilton, J. D. (2007). Convergence and divergence in the public dialogue on nuclear weapons cleanup. In B. Taylor, W. Kinsella, S. Depoe, & M. Metzler (Eds.), *Nuclear Legacies: Communication, Controversy, and the U.S. Nuclear Weapons Complex* (pp. 41–72). Lanham, MD: Lexington Books.
Hamilton, J. D., & Wills-Toker, C. (2006). Reconceptualizing dialogue in environmental public participation. *Policy Studies Journal, 34*, 755–775.
Haraway, D. (2008). *When Species Meet*. Minneapolis: University of Minnesota Press.
Hauser, G. (1999). *Vernacular Voices: The Rhetoric of Publics and Public Spheres*. Columbia: University of South Carolina Press.
Hendry, J. (2010). *Communication and the Natural World*. State College, PA: Strata.
Hope, D. (2004). The rhetoric of autobiography in women's autobiographical narratives: Lois Gibbs' *Love Canal: My Story* and Sandra Steingraber's *Living Downstream: An Ecologist Looks at Cancer and the Environment*. In S. L. Senecah (Ed.), *The Environmental Communication Yearbook*, Vol. 1 (pp. 77–98). Mahwah, NJ: Erlbaum.

Huspek, M., & Kendall, K. E. (1991). On withholding political voice: An analysis of the political vocabulary of a "nonpolitical" speech community. *Quarterly Journal of Speech, 77(1)*, 1–19.

Jensen, D. (2002/2004). *Listening to the Land: Conversations about Nature, Culture, and Eros.* White River Junction, CT: Chelsea Green Publish Co.

Killingsworth, M. J., & Palmer, J. S. (1992). *Ecospeak: Rhetoric and Environmental Politics in America.* Carbondale, IL: Southern Illinois University Press.

Klassen, J. A. (2011). Oiling the gears of public participation: The value of organisations in establishing the Trinity of Voice for communities impacted by the oil and gas industry. *Local Environment, 16*, 903–915.

Lange, J. I. (1990). Refusal to compromise: The case of Earth First! *Western Journal of Speech Communication, 54(4)*, 473–494.

Leopold, A. (1949). *A Sand County Almanac and Sketches Here and There.* London: Oxford University Press.

Lovelock, J. (1995). *The Ages of Gaia: A Biography of our Living Earth.* New York: Norton.

Marafiote, T., & Plec, E. (2006). From dualisms to dialogism: Hybridity in discourse about the natural world. In S. P. Depoe (Ed.), *The Environmental Communication Yearbook,* Vol. 3 (pp. 49–76). Mahwah, NJ: Erlbaum.

Meine, C. D. (1999). Leopold's voice: The reach of words. In C. D. Meine & R. L. Wright (Eds.), *The Essential Aldo Leopold: Questions and Commentaries* (pp. 314–330). Madison: University of Wisconsin Press.

Micheletti, M. (2003). *Political Virtue and Shopping: Individuals, Consumerism, and Collective Action.* New York: Palgrave MacMillan.

Milstein, T., & Krolokke, C. (2012). Transcorporeal tourism: Whales, fetuses, and the rupturing and reinscribing of cultural constraints. *Environmental Communication: A Journal of Nature and Culture, 6*, 82–100.

Mitra, A., & Watts, E. (2002). Theorizing cyberspace: The idea of voice applied to the internet discourse. *New Media and Society, 4(4)*, 479–496.

Muir, S. A., & Veenendall, T. L. (Eds.) (1996). *Earthtalk: Communication Empowerment for Environmental Action.* Westport, CT: Praeger.

Myerson, G., & Rydin, Y. (1996). *The Language of Environment: A New Rhetoric.* London: UCL Press.

Oravec, C. L. (1981). John Muir, Yosemite, and the sublime response: A study in the rhetoric of preservationism. *Quarterly Journal of Speech, 67*, 245–258.

Peeples, J. (2005). Aggressive mimicry: The rhetoric of wise use and the environmental movement. In S. L. Senecah (Ed.), *The Environmental Communication Yearbook,* Vol. 2 (pp. 1–18). Mahwah, NJ: Erlbaum.

Peeples, J., & DeLuca, K. M. (2006). The truth of the matter: Motherhood, community and environmental justice. *Women's Studies in Communication, 29*, 39–58.

Pezzullo, P. C. (2011). Contextualizing boycotts and buycotts: The impure politics of consumer-based advocacy in an age of global ecological crises. *Communication and Critical/Cultural Studies, 8*, 124–145.

Phillips, L., Carvalho, A., & Doyle, J. (Eds.) (2012). *Citizen Voices: Performing Public Participation in Science and Environment Communication.* Bristol, UK: Intellect.

Plec, E. (2013). Perspectives on human-animal communication. In E. Plec (Ed.), *Perspectives on Human-animal Communication: International Communication* (pp. 1–13). London: Routledge.

Plec, E., & Pettenger, M. (2012). Greenwashing consumption: The didactic framing of ExxonMobil's energy solutions. *Environmental Communication: A Journal of Nature and Culture, 6*, 459–476.

Rogers, R. A. (1998). Overcoming the objectification of nature in constitutive theories: Toward a transhuman, materialist theory of communication. *Western Journal of Communication, 62*, 244–272.

Roszak, T. (1992). *The Voice of the Earth*. New York: Simon & Schuster.

Rosteck, T., & Frentz, T. S. (2009). Myth and multiple readings in environmental rhetoric: The case of *An Inconvenient Truth*. *Quarterly Journal of Speech, 95*, 1–19.

Salvador, M., & Clarke, T. L. (2011). The weyekin principle: Toward an embodied critical rhetoric. *Environmental Communication: A Journal of Nature and Culture, 5*, 243–260.

Sandler, R., & Pezzullo, P. C. (2007). *Environmental Justice and Environmentalism: The Social Justice Challenge to the Environmental Movement*. Cambridge, MA: MIT Press.

Senecah, S. L. (2004). The trinity of voice: The role of practical theory in planning and evaluating the effectiveness of environmental participatory processes. In S. P. Depoe, J. W. Delicath, & M. A. Elsenbeer (Eds.), *Communication and Public Participation in Environmental Decision-making* (pp. 13–33). Albany: SUNY Press.

Shabecoff, P. (2003). *A Fierce Green Fire: The American Environmental Movement* (2nd ed.). Washington, DC: Island Press.

Shaiko, R. G. (1999). *Voices and Echoes for the Environment: Public Interest Representation in the 1990s and Beyond*. New York: Columbia University Press.

Singer, R. (2010). Neoliberal style, the American re-generation, and ecological jeremiad in Thomas Friedman's "Code Green." *Environmental Communication: A Journal of Nature and Culture, 4*, 135–151.

Stewart, C., Smith, C. A., & Denton, R. E. Jr. (2001). *Persuasion and Social Movements*. Prospect Heights, IL: Waveland Press, Inc.

Stillion Southard, B. A. (2007). Militancy, power, and identity: The silent sentinels as women fighting for political voice. *Rhetoric & Public Affairs, 10(3)*, 399–417.

Ullman, H. L. (1996). Thinking like a mountain: Persona, ethos, and judgment in American nature writing. In C. G. Herndl & S. C. Brown (Eds.), *Green Culture: Environmental Rhetoric in Contemporary America* (pp. 46–81). Madison: University of Wisconsin Press.

Waddell, C. (Ed.) (2000). *And No Birds Sang: Rhetorical Analyses of Rachel Carson's Silent Spring*. Carbondale: Southern Illinois University Press.

Walker, G. B., Senecah, S. L., & Daniels, S. E. (2006). From the forest to the river: Citizens' views of stakeholder engagement. *Human Ecology Review, 13*, 193–202.

Watts, E. (2001). "Voice" and "voicelessness" in rhetorical studies. *Quarterly Journal of Speech, 87(2)*, 179.

Watts, E. (2012). *Hearing the Hurt: Rhetoric, Aesthetics and Politics of the New Negro Movement*. Tuscaloosa: U. of Alabama.

Zeff, R. L., Love, M., & Stults, K. (Eds.) (1989). *Empowering Ourselves: Women and Toxics Organizing*. Falls Church, VA: Citizen's Clearinghouse for Hazardous Wastes.

Section I
Voice and Environmental Advocacy

1
Corporate Ventriloquism: Corporate Advocacy, the Coal Industry, and the Appropriation of Voice

Peter K. Bsumek, Jen Schneider, Steve Schwarze, and Jennifer Peeples

In the second decade of the 21st century, the U.S. coal industry is facing unprecedented challenges. While for many years coal provided nearly half of U.S. electricity, in the spring of 2012 that share dropped to below 40% and is expected to continue falling (Energy Information Administration, 2012).[1] Coal production is increasing not in Appalachia, the primary U.S. source for coal historically, but in Wyoming's Powder River Basin (Goodell, 2006). Market competition from the natural gas industry combined with well organized climate and anti-mountaintop removal (MTR) campaigns have significantly curtailed the production of new coal-fired power plants in the United States (EIA, 2012). Under the Obama administration, the Environmental Protection Agency appears to be somewhat more amenable than the Bush administration to regulating carbon emissions as a pollutant, and more interested in enforcing Clean Water Act provisions applicable to MTR mining (Broder, 2012). Combined with sharp reductions in the number of coal mining jobs due to the increased efficiency of coal mining techniques, these circumstances have put the coal industry in Appalachia in a precarious position.

The coal industry in Appalachia has responded to these circumstances by waging a multi-front corporate advocacy campaign. This campaign combines traditional tactics such as litigating, lobbying, and backing pro-coal candidates in local and national elections. But it also involves a series of sophisticated, coordinated public relations campaigns that seek to secure the hegemony of coal both regionally and nationally. Through trade associations and advocacy organizations that produce websites, advertisements, videos, and other messages, the campaigns seek to unify a range of people who are "speaking with one voice" about coal ("One Voice").

These campaigns and their creation of a "voice" for the coal industry are the focus of this chapter. Using theories of voice and appropriation, we argue that the coal industry's rhetoric operates through a process that we term *corporate ventriloquism*. In this rhetoric, the industry appropriates elements of neoliberal and neoconservative ideology and adapts them to the cultural circumstances specific to coal in Appalachia. It then "throws" this voice through "front groups" to create the impression of broadly based support for coal. Through corporate ventriloquism, the coal industry masks its own influence over the spaces and conditions for voice and undermines the value of dissenting, textured, and independent voices in public discussions about the future of coal.

We begin the essay by putting Nick Couldry's theory of "voice" under neoliberal regimes into conversation with rhetorical theories of appropriation to build the concept of corporate ventriloquism. We then map the complex array of organizations that enable the coal industry to speak as if it were a legitimate voice of the people. Next, we offer a two-part analysis of a "Faces of Coal" campaign, which is emblematic of the industry's use of corporate ventriloquism and its neoliberal commitments. Our conclusion draws out several implications about corporate ventriloquism and its relationship to voice, neoliberalism, and environmental controversy.

Neoliberalism and the crisis of voice

This chapter extends discussions of appropriation by moving from existing social movement analyses of strategy, tactics, terminology, and structure to focus on the use and manipulation of "voice" as an element of appropriation. Our consideration of voice relies on the work of media and communication theorist Couldry (2010), who theorizes a "crisis of voice under neoliberalism." Based on economic theories popularized by Friedrich von Hayek and Milton Friedman, neoliberalism is guided by the assumption that individual and political freedoms are dependent upon a political economic system of free markets, free trade, and strong private property rights (Harvey, 2005). Couldry positions neoliberalism as a discourse and an organizing rationality. According to Couldry, the "market-driven politics" of neoliberalism has undermined the regulatory powers of government and facilitated the expansion of market rationality into nearly all aspects of public and private life. Neoliberalism has led to the deregulation of markets and industries, the privatization and "marketization" of public services, and the decline of trade unions. With regard to environmental policy and regulation, marketization is characterized by the shift from "command and control" regulatory approaches to those based on "market incentives" such as "cap and trade" (see Hajer, 1995). The hegemony of neoliberalism extends beyond government policy by producing the cultural conditions that constrain *and* constitute subjectivity and agency in both the social and the political realms.

By focusing on voice, Couldry demonstrates how neoliberal rationality constrains and constitutes subjectivity and agency. As such, neoliberalism limits the possibilities of what can be said, frames political controversies as primarily economic in nature, and reproduces neoliberal ideology, like the idea of a free market, as "common sense." Couldry thus provides a normative theory of voice, which is offered as a counter-rationality to the hegemony of neoliberalism.

Couldry distinguishes two levels of voice: voice as process and voice as value. As process, voice is "the process of giving an account of one's life and its conditions," a chance to speak on one's own behalf (p. 7). Eric Watts (2001) notes that in rhetorical scholarship, this notion of voice is associated with "speaking" and informs critical projects designed to enable marginalized or alienated people to "find their own voice" (p. 182). According to Couldry, neoliberal rationality excludes and undermines the process of giving voice by excluding alternative viewpoints. This takes place when institutions fail to register individual experience, when they ignore collective views, and when societies are encouraged to believe that "voice need not be taken into account, because a higher value or rationality trumps them" (Couldry, 2010, p. 10). Under these conditions, the process of finding voice, unless it expresses market rationality, is rendered mute and moot.

This is why, for Couldry, voice means more than a chance to speak and be heard. It is not enough to give an account of one's life if the only rhetorical situations available are constrained by market rationality and its identities and values. Couldry's second level of voice is therefore voice as value. As a value, voice is "the act of valuing, and choosing to value, those frameworks for organizing human life and resources that *themselves* value voice [as a process]" (p. 2). Here Couldry is concerned with the way in which neoliberalism exerts influence over the conditions for voice. Neoliberal rationality, for Couldry, "provides principles for organizing action (in workplace, public services, fields of competition, public discussion) which are internalized as norms and values" (p. 12). Key among these norms and values are the association of freedom with the "entrepreneurial" self—the individual as a free and independent agent in a free market—and the devaluing and dismissal of forms of social solidarity such as trade unions.

Neoliberalism establishes paradoxical terms for voice, in other words. According to Couldry, neoliberalism seems to permit the *apparent* expansion of voice (say, through ever-expanding consumer choice), while voice is *in fact* limited to market expression. Individuals are offered ample opportunities to "voice" their opinion in the marketplace or using economic logic. Yet opportunities to express ideological commitments outside of market logics are increasingly scarce. At stake, then, is not only the creation of more opportunities for giving "an account of one's life," but also the types of *"values* [that can be] articulated through such voices" (p. 137).

In the remainder of this chapter, we explore this paradoxical nature of "neoliberal voice" by investigating how the corporate advocacy campaigns of the coal industry celebrate the process of voice—multiple, individual expressions of "self"—while simultaneously muting and dismissing those voices that articulate values counter to neoliberal ideology. This case study is, then, an extension of Couldry's (2010) project in that it attempts to uncover the neoliberal processes that obstruct the means of valuing voice. Rhetorical theories of appropriation can aid this extension by further unpacking the paradoxes of neoliberal voice.

Appropriation and corporate ventriloquism

Communication scholarship on coal industry information campaigns is limited. Some sociologists, however, have analyzed how the coal industry uses these campaigns to shape cultural understandings of coal mining and the coal industry within Appalachia. Bell and York (2010) note that "when there is a large scale-reduction in jobs, and employment no longer connects an industry to the community it pollutes," economic rationality cannot fully explain why communities continue to support that industry (p. 116).[2] In situations like this, they argue, other kinds of ideologies must bolster economic rationales, enabling companies to maintain their cultural and political dominance in the region. Similarly, Rebecca Scott (2010) discusses the way that coalfield residents "are constructed and construct themselves as coalfield residents and how the discursive structuring of their subjectivity shapes their environmental politics" (p. 17). Noting that "social analyses of mining are usually limited to economic and political fields," she argues that coal mining—and MTR in particular—is a "deeply cultural act, and the complex environmental politics of coal mining are, in part, struggles over the meaning of the practice," and that these meanings are further "enmeshed in networks of material signification" (p. 17).

These networks of material signification, which include ideas about private property, land ownership, gender, race and class commitments, and national identity, are an important rhetorical resource for the coal industry as it attempts to address its material decline (Scott, 2010, pp. 17–18). Another rhetorical resource utilized by natural resource industries facing organized opposition has been to modify their public persona as a means of popularizing their industry (e.g., Smerecnik & Renegar, 2010). One such approach has been to tap networks of signification by appropriating the powerful structures and/or discourses of other organizations in order to obtain, co-opt, or counter their influence or identity. Environmental communication scholars have identified four primary means of appropriation seen in environmental controversies: lateral appropriation, greenwashing, astroturf campaigns, and aggressive mimicry. To that list we add corporate ventriloquism.

The most benevolent of the four means of co-optation is *lateral appropriation*. Anspach, Coe, and Thurlow (2007) define lateral appropriation as "any instance in which means commonly associated with and/or perceived as belonging to one marginalized group are used by another marginalized group to further its own ends" (p. 100; see also Peeples, 2011). Lateral appropriation is an important tactic for groups who have limited material and symbolic resources (Anspach, Coe, & Thurlow, 2007). It is also used by powerful organizations, like corporations, to adapt hegemonic discourses to new circumstances. Unlike the other forms of co-optation, lateral appropriation does not attempt to challenge or undermine the discourse it appropriates. Rather, it extends it into new discursive fields.

The second form of appropriation is *greenwashing*, which Cox defines as "the act of misleading consumers regarding the environmental practices of a company or the environmental benefits of a product or service" (2010, p. 345; see also Shapiro (2004)). Pezzullo adds that greenwashing also includes "the deliberate disavowal of environmental effects" (2003, p. 246). As with whitewashing, the appropriation of environmental discourse is cosmetic, leaving the product, production, organizational structure, and/or corporate agenda intact. Cox (2010) describes three purposes of greenwashing: (1) product promotion, (2) organizational image enhancement, and (3) organizational image repair. The company name or logo is often the focal point, as the purpose of greenwashing is for the organization (at least superficially) to alter its public image.

Astroturfing, the third type of appropriation, refers to the "the controversial tactic of creating the illusion of a largely spontaneous grassroots protest that has in fact been organised by corporate-backed groups" (Murray, 2009).[3] Tactics can range from "public" letter-writing campaigns that are organized, paid for, and even written by companies, to establishing community organizations and NGOs, "front groups" that are directly or indirectly funded and managed by corporations. The purpose of this appropriation is to persuade potential constituents or decision-makers that the message comes from citizens who have vested interests in the outcome, as opposed to corporate beneficiaries who have difficulty engendering either the level of empathy or the rights that are given to "the people."

The final strategy of appropriation, *aggressive mimicry*, is similar to astroturfing, but takes appropriation one step further. An entity engaging in aggressive mimicry co-opts an opposing organization's structure and discourse in order to sow doubts about their opponent's identity, with the intended effect of distracting or destroying the opponent (Peeples, 2005). For Wise Use, a 1980s anti-environmental movement, it was claiming to be the "true environmentalists" (Peeples, 2005). Co-opting organizations are at times unable to make great legislative and political strides as they can be seen as inauthentic or false. But their power lies in forcing their opponent to defend its discourse, structure, and identity, thereby diverting time and limited resources from the mimicked movement's goals (Peeples, 2005).

Our analysis extends this discussion of co-optation by describing how the coal industry uses strategies of appropriation in ways that contribute to "the crisis of voice" identified by Couldry. Informed by the theories outlined above, we identify a practice that we have termed "corporate ventriloquism," which we define as the rhetorical process of corporations transmitting their voice through seemingly less powerful entities in ways that advance the interests of those corporations and undermine the value of voice in democratic processes. In the case of the Appalachian coal industry, corporate ventriloquism relies upon astroturfing to generate an alternative persona from which to speak. But it also laterally appropriates neoliberal market rationality, drawing on neoconservative discourses of family values and national security, to constitute a regional Appalachian identity and a national American identity that are dependent on coal.[4] In the same move, the neoliberal discourse calls into question any voice that opposes the hegemonic coal doctrine as "anti-American" or against Appalachian prosperity, thereby silencing any expression other than the one appropriated and approved voice of coal. Rather than merely decrying astroturf groups as fake or inauthentic, we advance the notion of corporate ventriloquism in order to help critics observe the ideological work accomplished by that process.

Astroturf: Mapping the friends, FACES, and voice of coal

Several industry-affiliated organizations promote the interests of coal in Appalachia. They frequently work together to organize rallies and protests, disseminate talking points via press releases and lobbying, and produce media messages and educational materials to advance industry positions. The most prominent of these organizations are the National Mining Association (NMA), a national trade organization whose primary mission is lobbying in Washington, DC, and the West Virginia Coal Association (WVCA), which coordinates pro-mining lobbying efforts at the state and regional level. These umbrella organizations also fund, or share funding with, a number of affiliate organizations, including Coal Mining Our Future, the Coalition for Mountaintop Mining, Citizens for Coal, the American Coalition for Clean Coal Energy, and Friends of America.

The two most visible campaigns of the WVCA are Friends of Coal (FOC), which launched in 2002, and Faces of Coal (FACES), which launched in 2009. According to the corporate watchdog website SourceWatch, the WVCA funds FOC, whose emphasis is primarily on improving the public relations and marketing environment for coal mining in Appalachia (SourceWatch, 2011). In turn, FOC funds the group FACES (Federation for American Coal, Energy and Security), which also serves as the home of the "Faces of Coal" group/campaign, whose mission is to underscore the economic and social dependence Appalachians have on coal mining (SourceWatch, 2009). While both groups address their campaigns to the Appalachian region, the FACES organization also addresses its campaigns to national audiences.

Both FOC and FACES represent themselves as grassroots groups. FOC claims to be run by volunteers; the FACES website states that it is made up of "an alliance of people from all walks of life who are joining forces to educate lawmakers and the general public about the importance of coal and coal mining to our local and national economies and to our nation's energy security" ("About Us"). Such language suggests that these campaigns are the result of local, homegrown efforts to promote coal mining. This suggestion is further supported by language that connotes "small-town" values and organization. For example, FOC describes itself as "an army of coal miners, their families, friends, neighbors, local and state business leaders, elected officials, doctors, lawyers, teachers, pizza delivery guys and students" (WVCA, 2011, p. 4).

A key element of the marketing strategy of both FOC and FACES, therefore, is to emphasize the ways in which coal is "us." This is a distinctly nostalgic vision of "us," which emphasizes conservative articulations of shared cultural and political values: family, free markets, and football. In essays, brochures, videos, event sponsorship, and baseball hat logos, the message of FOC and FACES is that coal is constitutive of Appalachian and, in many materials, American identity. According to one essay on the FOC website, "Coal is West Virginia! Coal is America!"[5] Such a statement stands in stark opposition to progressive arguments about "big coal" and corporate malfeasance (e.g., Goodell, 2006). According to the FOC/FACES narrative, if you are against coal, you are against "us," against America, against progress, against what we *do* (jobs), and against our way of life, which relies on cheap electricity produced by coal.

Although what counts as "grassroots" is contested (Cox, 2010), the fairly heavy-handed top-down funding and organizational structure, as well as a specific intent to make these campaigns *seem* as if they originated with everyday people and not from the industry itself, suggests astroturf. The idea that these are grassroots campaigns serves the rhetorical pairing of "coal" and "America," but does not necessarily reflect the origins of the campaigns.[6]

Corporate ventriloquism and the FACES of coal

FOC and FACES specifically target Appalachian audiences, but also are aiming to have national reach. To do so, the industry utilizes a two-pronged strategy. First, it builds "dummy" grassroots organizations through which it can "throw" its voice. This is the practice of ventriloquism. Next, it utilizes the persona of that grassroots organization to deliver its messages to targeted national audiences. Our analysis examines the neoliberal and neoconservative dimensions of coal's voice in these messages.

Constructing a neoliberal voice

The astroturf efforts of the Faces of Coal campaign became clearest in August 2009, when progressive bloggers discovered that images from the

FACES website—images of people who were literally meant to represent the "faces" of coal—were actually generic images from a service called iStockphoto, which supplies marketing campaigns with stock photos. The story, which was first posted at the progressive blog DeSmogBlog, quickly made the rounds of progressive blogs all over the web; within a week, bloggers at Treehugger, MSNBC's *The Rachel Maddow Show*, Grist, The Daily Kos, Appalachian Voices, and the Huffington Post had reported or reposted about the FACES debacle. Several bloggers interpreted the use of iStock photos as a sign of astroturf, a faking of grassroots support for the coal industry.

This interpretation was amplified when bloggers discovered that the FACES website was hosted by an organization called "Adfero," which was not a grassroots organization at all, but rather a K Street (lobbying) public relations firm. According to Jim Hoggan at DeSmog Blog, "Adfero doesn't specialize in spontaneous public advocacy. It specializes in crafting a 'custom-tailored message' and then recruiting 'key contacts' who can slam that message home" (Hoggan, 2009). For Hoggan, Adfero's involvement offered clear evidence that FACES was the product of "inside-the-beltway, fossil fuel funded conservative lobbyists." Within hours of this accusation, the attribution to Adfero had been removed from the FACES website, and the website was moved to a server in Michigan (Johnson, 2009). Other bloggers were quick to point out that the actual supporters (funders) of FACES were not listed on the website (Sheppard, 2009).

The FACES iStockphoto flap illustrates several important components of the coal industry's strategy. Primarily, the incident highlights astroturfing as a tactic that recognizes the rhetorical power of individual voices by making them *seem* present without actually being so. The "faces" that are intended to represent the inclusion of voice are actually empty signifiers of a dialogue that never takes place. In addition, the incident reveals the ways in which the industry attempts to "throw" its voice, to place its values and justifications literally in the mouths of individuals, albeit individuals who have no identity beyond that of surface representation.

Further, the incident reinforces how the industry's ventriloquism relies on appeals to economic identity, hailing its audience as consumers. The campaign directs attention away from coal extraction and pollution and toward coal consumption, an act which implicates everyone, and that can be rationalized and sanitized more easily. It is telling that none of the stock photos depicted miners or other coalfield workers. Rather, the "faces of coal" were shown participating in everyday activities—lifeguarding, playing high school football, whitewater rafting, standing in front of a flower business (Randolph, 2009). These individuals are connected to coal primarily through consumption, or as individuals whose idealized middle-class American lifestyles are (presumably) supported by coal.

This is a fundamentally neoliberal appropriation of voice: the "faces of coal" and the viewer of the campaign alike are called to interpret coal primarily through the market logic of consumerism. They "speak" to coal, about coal, or for coal as consumers. The FACES website did not attempt to provide a platform where coalfield workers and residents might "give an account of the conditions of their lives" (Couldry, 2010, p. 4). Instead, it attempted to hail its audience as "generic" model Americans by positioning them as electricity consumers, and asking them to identify with the way that coal enables their lifestyle.

Since the iStockphoto flap, the FACES website has been updated to include information about the organization's supporters, and it features what appear to be actual Appalachian residents who support the coal industry. The navigation menu has an "our supporters" tab that links to a "supporter quotes" page. This page contains a long series of photographs of individuals—some blue collar, but also many who look like doctors, nurses and other professionals—with quotes such as, "If it wasn't for coal our area would become a ghost town. Everything revolves around the mining industry. Coal is our present and our future." The quotes are attributed to supporters using a first name and last initial ("Supporter Quotes"). The page also has a "featured profile" button; at the time of this writing, it featured an artist who does oil paintings of Appalachian wildlife, accessible to her (it is implied) more readily after MTR mining:

> Sharon hopes that her oil paintings inspire Americans to come to West Virginia to see the natural beauty of the landscape and wildlife that is present in her hometown—and largely the result of mine reclamation. She wants to save the mining jobs in West Virginia, because if you save the mining jobs, you save the state.

Finally, there is a "supporters list" with associations and organizations that have donated to FACES, which was glaringly missing during the iStockphoto incident.

Two things are worth noting about this new and improved FACES website. First, although the organization now comes clean about its supporters, it is in many ways a more effective vehicle for corporate ventriloquism than it was before. This is because the "our supporters" tab flattens the relationship between individual supporters and corporate sponsors. The mix of elements on the "our supporters" tab treats the quoted individuals, local governmental organizations, and businesses associations (including the WVCA) as coequal "supporters" of FACES. It also claims over 60,000 individual supporters and lists a small sample by first name, last initial, and hometown. In effect, the website reduces large-scale and highly funded organizations, like the WVCA, to just one of many supporters listed alphabetically. This reinforces the description of FACES noted earlier, as "an alliance of people from all walks of life," whereby the WVCA (the organization that originally created FOC and FACES) becomes just another supporter. This flattening supports the claim that FACES is where everyone is given equal voice.

Flattening corporations so that they seem like individuals is another neoliberal strategy. Neoliberalism values individualism (with individuals defined as corporations *and* consumers) above all else because neoliberal rationality privileges market relationships over other forms of social organization (Couldry, 2010, p. 66). This rhetorical move is conducted masterfully by Don Blankenship, former CEO of Massey Energy Company, then the largest coal producer in West Virginia. In a 2010 public debate with Robert Kennedy, Jr. in Charleston, West Virginia, Blankenship no less than five times responded to Kennedy's attacks on "the industry" by hailing the audience *as* the industry:

> It is easy to say that the industry is evil. The industry is ruthless. The industry is destroying the climate and destroying the environment. [But] you are the industry. The people in this room, the people that are in the banks, the people that are working in the coal mines. The people—we're the ones that are making the decisions.... We are the industry. You are the industry. The people that are your neighbors and your teachers are the industry. So I don't know—again—what it is that we want to be so easily critical of "the industry" (in finger quotes). Because that is us.
>
> (Blankenship, 2010)

By arguing that the people are the "the industry," Blankenship effectively erases the disproportionately powerful role coal companies play in Appalachian society. In so doing, he throws the corporate voice onto and through the people, creating in one rhetorical move a unification between the industry and Appalachia, and an identity that is not merely pro-coal and pro-industry, but *is* coal and *is* industry. The framing of the Appalachian and American self as fundamentally corporate and consumerist is essential to undermining alternative voices. Ironically, Blankenship also provides his audience with the one form of social solidarity that is available to the neoliberal subject: a conglomerate of individuals organized for the purpose of defending industry and expanding free markets.

Appealing to neoconservative values

While reliant on astroturfing strategies, coal's corporate ventriloquism also utilizes the strategy of lateral appropriation. In the case of the coal industry, this means appropriating neoconservative discourses, which equate freedom with family and national security, to buttress neoliberal ideology. Through this process the coal industry utilizes the networks of material signification, identified by Scott (2010), as rhetorical resources to shape the meanings of coal. As it does so, it also attempts to reconstitute those networks in terms of neoliberalism.

The FACES airline ad we examine in this section appeared in airline magazines in December 2010.[7] The ad targets affluent frequent flyers, which the *Arbitron In-flight Media Study* describes as "a very select group of Americans" and as "successful professionals with sophisticated tastes and the income to pursue their interests" (Williams, 2006, p. 1). As such, it is not appealing to a narrow Appalachian identity, and the dominant feature of the ad—an image of a young boy running with an American flag in a pasture—seems to have absolutely nothing to do with coal or electricity. Instead, the image is a generic reference to America, idealized American family life, and idealized American citizenship. This appropriation of neoconservative themes, both in the image and in the accompanying text, reveals how the coal industry attempts to create a consubstantial relationship between the coal industry, the market economy, and the nation. Further, it reveals how the coal industry's corporate ventriloquism conflates diverse national voices and viewpoints into "one voice."

The FACES airline ad specifically valorizes the "heartland" virtues that are prominent in American culture and politics. The ad's boldest visual element is a photograph of an American flag, held upright by a child. The reader symbolically enters the image from below, as if seated on a lawn chair at the edge of the intimate scene, and looks up with the child toward the flag. The intimate framing connotes family, and the child unmistakably functions as a signifier of hetero-normative family values and whiteness. However, this is not only a reference to "family values"; the ad articulates those values as American citizenship. Rebecca Scott suggests that in America the idealized rural citizen is characterized as "independent, brave, honest, and most important, always ready to sacrifice for the good of the nation" (2010, p. 37). America sustains and protects the family (the flag unfurls above the boy). The family sustains America by acting as the keeper of its virtue, and by serving its interests (the boy holds up the flag). Within this set of symbolic associations, the idealized American family is constituted as the bedrock of America.

The pastoral signifiers in the image are just as unmistakable—the red barn, pasture fencing, and mountains in the distance all connote the tranquility, peace, and innocence that are referenced in both British and American Romantic traditions (Williams, 1973; Garrard, 2004). As Raymond Williams notes, country life has often represented "an innocent alternative to ambition, disturbance and war" (Williams, 1973, p. 24). The FACES ad thus constructs a quintessentially American scene, calling to mind the innocence, purity, and Christian piety associated with idealized rural life in America, a common theme in Appalachian rhetoric about coal (Scott, 2010).

Although the image is symbolically important, the ad also features a long text box where we learn the ad is in fact for coal. The bolded headline of the ad reads, "American Power depends on American Coal." Power, in this statement, takes on two meanings; it stands for energy as well as international military, economic, and cultural hegemony. This text also reinforces the visual ideograph of the flag, and parallels the visual imagery of dependence: just as the nation depends on the family, so too does the nation depend on coal. A large blank space separates the headline from a list of four couplets, which supply the reader with evidence for the claim of dependency on coal. The space encourages the reader's eye to wander back to the image of the boy with the flag reinforcing the connection between the image and the claim "American Power Depends on American Coal." The eye then returns to the factoid couplets.

The couplets are arranged into three sets of two factoids that support the claim of American dependence, and are followed by a final couplet, which first threatens the reader, and then offers reassurance:

> Coal is America's most abundant energy source.
> Coal is America's most affordable energy source.
> Coal provides nearly 50% of America's electricity needs.
> Coal keeps that electricity affordable for millions.
> America needs jobs.
> America needs economic growth.
> That won't happen without electricity generated from coal.
> **Keep the Lights On, America.**

The first two couplets are arranged with parallel construction, each sentence beginning with the term "coal," while the third couplet begins with the term "America." This structure further defines and reiterates the relationship between coal and America: each depends on the other in a syllogistic or constitutive manner. America needs coal to be "powerful"; coal needs America to use its "power."

The first couplet stresses coal as an "energy source," and its first sentence is a clear reference to energy independence, encouraging the reader to think of problems associated with foreign sources of energy such as war in the Middle East and price spikes. The comfort and innocence portrayed in the image is now threatened. The second sentence on affordability plays to both individual consumer interests and the importance of an affordable energy source for geopolitical power. The term "energy source" is ripped from the pages of policy white papers and the punditry of Sunday morning talk shows. This couplet defines American power in terms of geopolitical power.

The second couplet emphasizes "electricity" and suggests that what is good for the nation is good for the individual consumer. The first sentence, "Coal provides nearly 50% of America's electricity needs," emphasizes the beneficial relationship between coal and the nation and suggests that the electricity that powers your family's home is likely generated by coal. The second sentence repeats the affordability theme in the first couplet and emphasizes the direct benefit of coal to the individual consumer—your electricity bill is "affordable" because of coal, while also suggesting that affordable electricity is good for the nation's economy. This point is further emphasized in the third couplet.

The third couplet provides general statements about what "America needs" that warrant the reader's support of coal. Namely, America needs "jobs" and "economic growth." The first sentence does double duty. It first anticipates environmental controversy associated with coal in the overly simplified frame of "jobs vs. environment." This frame dismisses environmental concerns associated with coal, erases the complexities associated with declining jobs in coal country, and narrows the economic frame within a neoliberal context. It also functions as a trope that reiterates the connotations in the image associated with heteronormative family. "Jobs" in this case, and in this proximity to the image in the ad, is a clear reference to being a family provider and a breadwinner. Thus, the reference to jobs in the airline ad is not just an economic argument, but also a morality claim. American family values depend on jobs. The ad builds on this theme in the next sentence, "America needs economic growth." Here the reference is not to the jobs that underpin the morality and virtue of the nation, but to the link between economic growth and national security. Family security depends on national security.

In the final couplet, parallel construction is abandoned in favor of a threat followed by a reassuring slogan. The first sentence, "That won't happen without electricity generated from coal," is the threat. At first glance, it asserts that jobs and economic growth will be jeopardized "without electricity generated from coal." However, when coupled with the associations of family values with jobs and geopolitical power with economic growth, the ad goes beyond the simple claim of jobs vs. the environment. Rather, the ad as a whole implies that the entire American way of life is threatened, and even being attacked. The last line in the couplet is then set apart from the previous seven lines and bolded: "**Keep the Lights On, America**." This command clearly echoes the timeworn slogan from billboards throughout Appalachia, "Coal Keeps the Lights On." In this context and in relation to this image, it seems bizarrely out of place. The image of the boy and flag is completely devoid of any references to lights; however, the reference to "keeping the lights on" clearly connotes home and family life. The old saying "We'll keep the lights on for you" is evoked.

The FACES ad is a quintessential example of how the coal industry and its proxies strategically appropriate "one voice" to rhetorically finesse material contradictions generated by neoliberal ideology. By drawing upon ideographs (Family, Security, Prosperity), commonplace tropes (pastoral, heartland), and the metaphors of neoliberal ideology (market competition, economic growth, power) the ad positions its audience as atomized, self-sufficient individuals—breadwinners and caretakers who have "earned" independence and are responsible for providing their families with security. Coal is simultaneously represented as a consumer good (an "abundant" and "affordable energy source") upon which the family depends, and as the resource upon which the nation depends for "power." The reference to "power" in this case does triple duty, referring to the electricity that powers our homes, the cheap electricity and the jobs created by the coal industry that power the nation's economy, and the resultant geopolitical power associated with a strong economy and independence from foreign sources of energy. Power comes from the seamless unity of the social, economic, and political realms, but economic growth fueled by coal plays the pre-eminent role.

Ultimately, the ad reinforces the neoliberal orientation toward coal found on the FACES website by bolstering it with a powerful appeal to a neoconservative American identity. The audience is asked not only to identify with coal as consumers and patriots, but also to embrace their dependence and reliance upon the coal industry. In other words, the relationship between the household breadwinners and caregivers and the coal industry is analogous to the relationship between the boy in the image and the breadwinners and caregivers who are asked to identify with him as they sit in the seat of an airplane.

This helps explain why FACES chose to place the ad in airline magazines. The cultural situation in the United States post-9/11, and the unique hyper-reliance of airline travelers upon the security-industrial complex, make the relationship of dependent freedom, of making sacrifices for freedom, a familiar one. By articulating coal to the values of family, freedom, and security, the ad implies that any negative aspects of coal extraction, production, or consumption are simply the price of freedom, the sacrifices we make for "our way of life."

Corporate ventriloquism as the voice of neoliberalism

Our analysis has described corporate ventriloquism as the key process by which the coal industry has attempted to negotiate a range of economic, cultural, and ideological challenges. Through strategies such as astroturfing of grassroots organizations and lateral appropriation of neoconservative discourse, the industry has crafted a voice through which it articulates coal production and coal-generated energy with economic prosperity and neoconservative values of family, nation, and security. Rhetorically, this articulation ultimately advances a neoliberal ideology that is conducive to the interests of the coal industry, but problematic for both the process and value of voice in public discourse surrounding coal.

Thus, we have aimed to demonstrate that corporate ventriloquism provides a useful entry point for examining the tension between voice and neoliberal ideology as identified by Couldry. In the remainder of this essay, we discuss some of the specific implications our analysis has for understanding Couldry's crisis of voice, as well as the role of voice in environmental controversies.

First, **corporate ventriloquism aids neoliberalism by consistently positioning audiences as market participants, thereby eliding the difference between the expression of voice and the functioning of markets.** As astroturf groups reframe industry interests by exclusively depicting the jobs, lifestyles, and everyday practices that are supported through corporate activity, they constitute audience members primarily as market participants. In our case study, local residents are invited to celebrate their economic connection to the coal industry, while the far-flung audiences of the FACES ads are positioned as consumers desiring affordable energy and workers desiring jobs and economic growth. The gendered, familial, and national dimensions of identity serve to reinforce one's roles in the market. The lateral appropriation of neoconservative imagery and tropes, along with an economic framing that constructs considerations of ecological degradation or social injustice as the price of freedom in a competitive global market, forecloses voice by celebrating "the citizenry's presumable essential socioeconomic solidarity to the exclusion of its constitutive political differences" (Vivian, 2006, p. 4). Other forms of identity-making or articulations of voice which might critique or lie outside of market activity are noticeably absent or foreclosed.

Second, **corporate ventriloquism enables industries to celebrate *and* undermine voice in environmental controversies**. Couldry notes that in democratic contexts we can often identify "rationalities that *do not* directly deny the value of voice outright (indeed, in some contexts they may celebrate it), but work in other ways to undermine the provision of voice at various levels" (2010, p. 10). The advocacy campaigns of FOC and FACES provide an exemplar for the contradictory and at times paradoxical character of voice in neoliberalism. On one hand, these campaigns appear to valorize the inclusion of individual voices through personal photos and quotations. The faces of "real" people on the FACES website become models for members. On the other hand, the rhetorical resources appropriated by coal do not enable individuals to provide "an account of one's life and its conditions" beyond generic yet powerful themes of family, patriotism, and consumerism that reinscribe neoliberal ideology. These rhetorical resources arguably undermine the value of voice in Couldry's sense by constraining individual accounts within the neoliberal presumption that "market functioning is the privileged reference-point" for all other modes of social organization (2010, p. 23).

Third, **corporate ventriloquism may become the preferred modality of voice under neoliberalism, to the extent that it obscures the fundamental tension between voice and neoliberal ideology**. The simultaneous celebration and undermining of voice noted above reveals a key moment in negotiating the contradictions of neoliberalism. While the existence of astroturf groups implicitly acknowledges that voice matters, neoliberalism devalues voice relative to effective market functioning. Thus, neoliberal ideology must incite a diverse array of market-oriented voices in order to negotiate "the tension between neoliberal doctrine and the value of voice" (Couldry, 2010, p. 11). Put differently, corporate ventriloquism is a way of recognizing voice under conditions of neoliberalism—but it is a voice that, in reinscribing neoliberal ideology, is not really a voice at all.

Fourth, **resistance to corporate ventriloquism requires more than identification of the "fake" character of astroturf groups.** Merely identifying groups as corporate-funded forms of astroturf is a necessary, but insufficient step for activists and critics who wish to thwart the dominance of corporate interests. Beyond identifying the instrumental creation of "fake" grassroots groups, our analysis of corporate ventriloquism explains how astroturf groups rhetorically hail an audience around a set of "real" interests and shared values, unifying all those who see themselves as connected with an industry. It is this deeper cultural alignment around the neoliberal equation of market functioning with the public interest that must be interrogated in order to resist the operations of corporate ventriloquism.

Finally, **corporate ventriloquism aids neoliberalism by effacing the differences between individuals and corporations.** This is another way that corporate ventriloquism differs from astroturfing. While astroturfing characterizes industry-supported organizations as spontaneous, grassroots collections of interested citizens, corporate ventriloquism goes one step further to characterize corporate interests as on par with citizen participants in the economic and political milieu. In our case study, rather than hide their identity as supporters, companies and trade associations reveal their connection to astroturf organizations, but then contextualize that connection by placing themselves alongside other individual supporters. This flattening obscures disparities in financial and political power wielded by different participants in the coal-industrial complex, making all voices appear equal. It also reinforces the legal construction of the corporation as an individual. This implication is especially significant in light of recent legal decisions such as *Citizens United v. Federal Election Commission*, which expanded First Amendment rights to corporations and dramatically altered the political economy of voice in the United States.

The juridical trick of defining corporations as individuals illustrates the hegemony of neoliberalism. The advantage now enjoyed by corporations to spend money to influence elections, combined with their already sizable financial advantage with regard to influencing public policy through lobbying and litigation, creates a playing field that is not just uneven, but rigged, putting advocates of alternative perspectives at a significant disadvantage. This suggests that the political, legal, and economic structures that enable this advantage must become a focus of scholarly criticism and environmental advocacy.

Matt Wasson (2012), director of programs for the environmental advocacy organization Appalachian Voices, provides a cogent example of this problem in his analysis of 2012 presidential election results in Appalachian, where pro-coal candidates increased their share of the vote in comparison to the 2008 election. Wasson argues that not only have national environmental advocates long ignored coal communities, focusing their attention on climate change and shutting down coal-fired power plants, but that local groups are simply unable to compete with the resources of the coal industry and its allies:

> "There are groups like Kentuckians for the Commonwealth that are doing extraordinarily effective organizing in regions where coal is mined, but when a group like Americans for Prosperity comes in with an $11 million ad campaign and bottomless pockets for on-the-ground organizing, we're in the position of bringing a knife to a nuclear showdown."
>
> (Wasson, 2012)

Wasson suggests that mainstream environmental organizations should pay more attention to organizing and promoting alternative economic opportunities in Appalachia. Further, he notes that "mountain top removal and drinking water pollution are potent 'gateway issues' that have inspired many residents to question the honesty and benevolence of the coal industry and their political allies" (Wasson, 2012). Advocacy organizations like Appalachian Voices and Kentuckians for the Commonwealth have been utilizing these gateway issues to broaden the scope of their resistance campaigns, making connections between environmental injustice and the neoliberal distortion of corporate power. However, in a region where the coal industry is king, and in a country where the deck is increasingly stacked (both ideologically and structurally) in favor of corporations, it is difficult to imagine that this process of voice can be heard in the absence of a renewed valuing of voice.

Conclusion

Corporate ventriloquism should thus be seen as a significant contributor to "the crisis of voice under neoliberalism." The coal industry's use of corporate ventriloquism may be a harbinger of things to come under neoliberal "democracy," which distorts democratic practices to the extent that it forecloses public participation, corrodes social ties, and privileges free markets above all else. By exposing the paradoxical nature of neoliberal voice and calling attention to its implications, we have endeavored to demonstrate that corporate ventriloquism is more complex than astroturf. In utilizing corporate ventriloquism, corporations do not attempt to hide behind "front groups"; rather, they construct a corporate voice that is positioned as the voice of citizenship. This voice is itself the ultimate expression of neoliberal ideology. In turn, this corporate voice serves as a powerful persona for laterally appropriating, adapting, and articulating rationales for neoliberalism. The concept of corporate ventriloquism enables us to focus both on the way corporations laterally appropriate such rationales, and on the way they construct their voice.

The matter of voice and the question of whether corporations can or should speak for us, or as one of us, are themselves issues to which communication scholars and citizens in general must give voice. Foregrounding the importance of voice, and reimagining social organizing principles around the importance of valuing voice, offer one way to connect advocacy with critiques of the ideological and rhetorical rationales that enable structural imbalances in the political economy of voice. Indeed, communication scholars are uniquely positioned to comment on the importance of balancing the advocacy playing field.

Notes

1. Though there are a number of factors leading to this precipitous decline, the bulk of the decrease is largely due to an increasing reliance on unconventional natural gas produced from hydrofracturing (fracking) shale plays.
2. For example, the number of coal miners employed in West Virginia has decreased from over 125,000 in 1945 to less than 25,000 in 2005. See Bell and York (2010, pp. 114–115). Shirley Stewart Burns' 2007 book *Bringing Down the Mountains* details the social, economic, and environmental costs of such shifts to residents of Appalachia.
3. For a discussion of corporate astroturfing and climate change see the *New York Times* editorial "Another astroturf campaign" (2009).
4. David Harvey (2005) situates neoconservative discourse as an extension of neoliberalism. Nationalism, militarism, and conservative "family values" are mobilized to sustain neoliberalism.
5. http://www.wvcoal.com/docs/Coal%20Facts%202011.pdf
6. Measuring the actual impact of these campaigns is beyond the scope of this paper; however, FOC touts its success on its own website, where it states:

> Before the FOC campaign began, surveys indicated that many people in the state had concerns about the coal industry and its role in the state.

A study taken just a couple of years ago indicated that most people today trust the industry (65 percent), up by some 17 percent since 2002.

("FOC Bowl," 2009)

Similarly, according to the marketing firm Preston-Osborne, which handles the FACES campaign in Kentucky, "Compared to a baseline survey conducted in April 2009, there was a notable drop in the percentage of respondents who strongly agree that mountain top coal mining should be banned in Kentucky" (http://preston-osborne.com/portfolio/faces-of-coal-3/).
7. See http://www.facesofcoal.org/index.php?photos&gallery_id=2. It is possible that this print advertisement appeared in other publications. We do know that it appeared in airline magazines in December of 2010. On the FACES website a pdf version of the ad is labeled "Airline Ad." If the ad was published elsewhere it likely targeted similar affluent audiences. The image in the ad was also used in a poster, labeled on the FACES website as "WV Fair Ad," with different text.

References

"About Us." (n.d.). Federation for American Coal, Energy and Security. Retrieved 3 June 2012 from http://www.facesofcoal.org/index.php?about-us
Another astroturf campaign [editorial] (4 September 2009). *New York Times*, p. 20.
Anspach, W., Coe, K., & Thurlow, C. (2007). The other closet? Atheists, homosexuals and the lateral appropriation of discursive capital. *Critical Discourse Studies, 4(1)*, 95–119. doi:10.1080/17405900601149509
Bell, A. E., &York, R. (2010). Community economic identity: The coal industry and ideology construction in West Virginia. *Rural Sociology, 75(1)*, 111–143.
Blankenship, D. (2010). Kennedy vs. Blankenship climate debate, Part 7/9, 8:48. Retrieved 4 June 2012 from http://www.youtube.com/watch
Broder, J. (2012). Court reverses E.P.A. on big mining project. *New York Times*, 23 March. Retrieved 4 June 2012 from http://www.nytimes.com/2012/03/24/science/earth/court-reverses-epa-saying-big-mining-project-can-proceed.html
Burns, S. S. (2007). *Bringing Down the Mountains: The Impact of Mountaintop Removal on Southwest Virginia Communities*. Morgantown, WV: West Virginia University Press.
Couldry, N. (2010). *Why Voice Matters: Culture and Politics after Neoliberalism*. Thousand Oaks, CA: Sage Publications.
Cox, J. R. (2010). *Environmental Communication and the Public Sphere* (2nd ed.). Thousand Oaks, CA: Sage Publications.
Energy Information Administration (2012, May). Short-term energy outlook. Retrieved 3 June 2012 from http://www.eia.gov/forecasts/steo/pdf/steo_full.pdf
"FOC Bowl." (2009). Friends of Coal. Retrieved 3 June 2012 from http://www.friendsofcoal.org/foc-bowl.html
Garrard, G. (2004). *Ecocriticism*. London: Routledge.
Goodell, J. (2006). *Big Coal: The Dirty Secret Behind America's Energy Future*. New York: Houghton Mifflin.
Hajer, M. A. (1995). *The Politics of Environmental Discourse: Ecological Modernization and the Policy Process*. Oxford: Clarendon Press.
Harvey, D. (2005). *A Brief History of Neoliberalism*. Oxford: Oxford University Press.

Hoggan, J. (2009). New "grassroots" pro-coal group backed by K-Street PR firm. *DeSmogBlog*. Retrieved 3 June 2012 from http://www.desmogblog.com/new-grassroots-pro-coal-group-backed-k-street-pr-firm

Johnson, B. (2009). The real FACES of coal: Adfero's shadowy GOP beltway astroturf operatives. *ThinkProgress Blog*. Retrieved 3 June 2012 from http://thinkprogress.org/climate/2009/08/27/174412/faces-of-adfero/?mobile=nc

Murray, J. (2009, August 17). Greenpeace uncovers "astroturf" campaign to challenge U.S. climate bill. *Green Business: Sustainable Living*. Retrieved 3 June 2012 from http://www.businessgreen.com/bg/news/1806738/greenpeace-uncovers-astroturfcampaign-challenge-us-climate

"One Voice." (2010). Friends of Coal. Retrieved 3 June 2012 from http://www.friendsofcoal.org/the-friends-of-coal-speaking-with-one-voice.html

Peeples, J. A. (2005). Aggressive mimicry: Wise use and the environmental movement. In S. Senecah (Ed.), *Environmental Communication Yearbook*, Vol. 2 (pp. 1–18). Mahwah, NJ: Lawrence Erlbaum Associates.

Peeples, J. A. (2011). Downwind: Articulation and appropriation of social movementdiscourse. *Southern Communication Journal, 76(3)*, 248–263.

Pezzullo, P. C. (2003). Resisting "national breast cancer awareness month": The rhetoric of counterpublics and their cultural performances. *Quarterly Journal of Speech, 89(4)*, 345–365. doi:10:80/0033563032000160981

Randolph, J. W. (2009). FACES of coal are iStockphotos?! *Grist*, 27 August. Retrieved 4 June 2012 from http://grist.org/article/2009-08-27-faces-of-coal-are-istockphotos/

Scott, R. A. (2010). *Removing Mountains: Extracting Nature and Identity in the Appalachian Coalfields*. Minneapolis, MN: University of Minnesota Press.

Shapiro, E. (2004). Greenwashing! *Communication Arts, 46(7)*, 198–205.

Sheppard, K. (2009). Who are the faces behind FACES of coal? *Grist*. Retrieved 3 June 2012 from http://grist.org/article/2009-08-20-who-are-the-faces-behind-faces-for-coal/

Smerecnik, K. R., & Renegar, V. R. (2010). Capitalistic agency: The rhetoric of BP's Helios power campaign. *Environmental Communication: A Journal of Nature and Culture, 4(2)*, 152–171.

SourceWatch (2009). Federation for American Coal, Energy and Security. Last modified 27 August 2009. Retrieved 3 June 2012 from http://www.sourcewatch.org/index.php?title=Federation_for_American_Coal,_Energy_and_Security

SourceWatch (2011). Friends of Coal. Last modified 8 February 2011. Retrieved 3 June 2012 from http://www.sourcewatch.org/index.php?title=Friends_of_Coal

"Supporter Quotes." (n.d.) Federation for American Coal, Energy and Security. Retrieved 3 June 2012 from http://www.facesofcoal.org/index.php?supporter-quotes

Vivian, B. (2006). Neoliberal epideictic: Rhetorical form and commemorative politics on September 11, 2002. *Quarterly Journal of Speech, 92(1)*, 1–26.

Wasson, M. (2012). Some electoral math for "all you climate people." *Think Progress*. Retrieved 7 March 2013 from http://thinkprogress.org/climate/2012/11/21/1227771/some-electoral-math-for-all-you-climate-people/

Watts, E. K. (2001). "Voice" and "voicelessness" in rhetorical studies. *Quarterly Journal of Speech, 87(2)*, 179–196.

West Virginia Coal Association (2011). West Virginia Coal: Fueling an American Renaissance. Retrieved 3 June 2012 from http://www.wvcoal.com/docs/Coal%20Facts%202011.pdf

Williams, D. (2006). The arbitron in-flight media study: Exploring frequent flyers' engagement with airline magazines and in-flight TV. *Arbitron*. Retrieved 4 June 2012 from www.arbitron.com/downloads/in-flight_media_study.pdf

Williams, R. (1973). *The Country and the City*. New York: Oxford University Press.

2
Defending the Fort: Michael Crichton, Pulp Fiction, and Green Conspiracy

Patrick Belanger

This chapter addresses the intersection of voice, narrative, and argument. All seek an audience, and all aim to constitute evaluative grounds. My case study is science-fiction author Michael Crichton's broad critique of the North American environmental movement as exemplified in two texts: a 2003 speech titled "Environmentalism as religion," and the 2004 thriller novel *State of Fear*. Each work strategically orchestrates the spheres of personal, technical, and public argumentation (Goodnight, 1982) and thus invites critique of the politically dexterous voice of a prominent writer.

Crichton's rhetorical success derives largely from his capacity to assert credibility by orienting the grounds of argument to his advantage. I argue that he does so through the tactical employment of voice. Through a patronizing voice, Crichton assails environmental thought as naïve and/or faith-based, and thereby relegates such thought to the private (apolitical) sphere. Through a populist voice, he transforms the grounds for discussion about environmental science from the technical to the public sphere and thereby authorizes his own (egalitarian) participation. Whereas the patronizing voice depicts environmentalists as misguided, the populist voice indicts such individuals as intellectually and ethically corrupt and thus dangerous. This move situates Crichton as paragon of both scientific integrity and common sense, and allows him to disregard the existing scientific literature. Asserting interpretive authority, he situates himself as public intellectual and defender of objective rigor.

In sum, I argue that Crichton employs two forms of voice to persuade audiences that environmental concerns (such as climate change) are fabrications perpetuated by dogmatic and/or radical environmental thought. As an international bestseller, *State of Fear* commands an audience far beyond those already positioned to respect Crichton's conservative message. The novel's audience thus consists of both casual science-fiction fans (from multiple political orientations) and environmental skeptics. The 2003 speech, by contrast, is addressed more clearly to the latter audience: individuals averse to "radical" environmental ideologies and sympathetic to the idea that bureaucratic corruption has distorted environmental policy. While the patronizing and populist voices seem contradictory (the first champions expertise, the second undermines the world's leading experts), Crichton's rhetorical strategy requires both moves. Due to his high profile, Crichton's arguments achieve broad publicity (including to unsympathetic audiences). This wide audience compounds the difficulties inherent in challenging the scientific establishment at its own game. Together, the patronizing and populist voices allow him to disregard the scientific literature, and do so without ceding the grounds of objectivity—he simply (and disingenuously) claims unique authority in that arena. At heart, Crichton asserts credibility by encompassing scientific concerns through a style and tactical ethos analogous to John McCain's 2000 presidential campaign bus: the "straight-talk express." His implicit argument proceeds: trust me—I will translate science into language you'll understand. Ultimately, by genre-switching between the technical, public, and personal spheres of argument, Crichton establishes the grounds of evaluation to his advantage.

Crichton claims to walk two roads: as popular author and as scientist. Yet only in the realm of the former can he validly profess expertise. His official website notes that he earned an M.D. from Harvard Medical School, but then defines him as "a writer and filmmaker, best known as the author of *Jurassic Park* and the creator of *ER*" and as "one of the most popular writers in the world [whose] books have been translated into thirty-eight languages, and thirteen...made into films" (About Michael Crichton, 2012). Unquestionably successful as an author, his list of professional work consists exclusively of contributions to the entertainment industry in the forms of novels, television shows, films, and video games. He has never practiced medicine. Credibility is not migratory. The central question is: why does Crichton (an expert science-fiction author) find a sympathetic audience in an arena beyond his area of competence when others with genuine expertise in environmental science do not? The answer extends beyond celebrity status—many celebrities have lent their voices to political, commercial, or social causes with only limited success. Prominent examples of failed celebrity campaigns are found in politics (Norman Mailer, Garry Kasparov, Shirley Temple, and P.T. Barnum), and advertising (Tiger Woods, Michael Vick). While celebrities generally draw enhanced media attention, the results of that exposure are variable (Rawlings, 2010).

Fusions of science fiction with policy recommendations are relevant only to the degree that they achieve an audience. In this respect, Crichton is an unqualified success. His vast audience extends to the corridors of United States federal power. The year after *State of Fear*'s publication, self-confessed fan Senator James Inhofe (who has pronounced global warming to be "the greatest hoax ever perpetrated on the American people") invited Crichton to Washington to counsel the United States Senate on the issue of climate change. *State of Fear* was required reading for the Senate Committee on Environment and Public Works (Janofsky, 2005). Although he received an unfavorable reception from certain lawmakers, the simple fact that he was presented with such an opportunity indicates that there are prominent lawmakers who welcome his views. Beyond this elite arena, his audience consists of millions more. He has sold over 200 million books, won an Emmy, a Peabody, and a Writers Guild of America award. Between 2007 and 2012 *State of Fear* was the tenth bestselling novel at United States airports (Time.com, 2012). Delivered at the Commonwealth Club in San Francisco, Crichton's "Environmentalism as religion" speech also enjoyed a high profile. Founded in 1903 as a national forum on public issues, the Club has hosted an array of prestigious speakers. Crichton joined a group of alumni including Teddy Roosevelt, Martin Luther King, Jr., Ronald Reagan, Bill Clinton, Arnold Schwarzenegger, Nancy Pelosi, and Bill Gates.

Crichton's creative work often plays upon public fears of modern institutions and a "millennial theme of techno-dystopia" (Goodnight, 1995, p. 279). The two texts here analyzed employ a novel slant on the genre. As always, hubris is the spark that leads to downfall—yet the antagonists are not mechanistic scientists, but rather fanatical environmentalists. Yet this is no easy target; environmentalism (broadly defined) enjoys widespread support among the United States public. A 2011 survey of 1010 adults conducted by the Yale Project on Climate Change Communication determined that approximately 64% of Americans support protecting the environment even at the expense of economic growth (Leiserowitz, Maibach, Roser-Renouf, & Smith, 2011, p. 33). Crichton's awareness of this cultural trend is indicated by his website's tactical apology that, with regard to climate science, "his conclusions have been widely misstated" ("About Michael Crichton," 2012). However this is but one phrase within a much larger body of work.

Before proceeding, it may be useful to clarify that Crichton's views on climate change contradict the conclusions of major international scientific organizations such as the Intergovernmental Panel on Climate Change (IPCC). The IPCC is an international authority on climate science and joint recipient of the 2007 Nobel Peace Prize. Established in 1988 by the World Meteorological Organization and the United Nations Environment Programme, it is charged with assessing "the scientific, technical and socioeconomic information relevant for the understanding of the risk of human-induced climate change" ("IPCC History," 2012). The organization publishes a report every five years, and these documents, summarizing the work of several hundred scientists from more than 100 countries, incorporate all relevant findings from within the past five years. IPCC reports are policy relevant but not policy prescriptive; they are designed to inform international political negotiations regarding both the threats posed by, and realistic response strategies to, climate change. In contrast to Crichton's assertion that "the evidence for global warming is far weaker than its proponents would ever admit" (2003, p. 18), the IPCC's 2007 report confirms that such warming is occurring and asserts that "Most of the observed increase in global average temperatures since the mid-20th century is *very likely* due to the observed increase in anthropogenic greenhouse gas concentrations" (emphasis in original) (Solomon et al., 2007, p. 10). Here the phrase "very likely" refers to expert judgment of an outcome probability exceeding 90% (p. 3). These findings are corroborated by the World Bank (2012), the United States Department of State (2009), the Royal Society (United Kingdom and the Commonwealth) (2012), and the National Academy of Sciences (Forest & Feder, 2011).

In *State of Fear's* first appendix, Crichton counters these august organizations in straightforward prose: "In science, the old men are usually wrong. But in politics, the old men are wise, counsel caution, and in the end are often right" (2004, p. 580). It is not my intention to evaluate the contemporary climate science literature. Regardless of IPCC scientists' ages and sexes, I interpret the organization's conclusions as authoritative, and then concentrate on Crichton's tactical employment of voice to mobilize doubt in the face of empirical evidence.

Voice and credibility

The term "voice" is somewhat elusive. It may signify a wave-based auditory phenomenon (sound), expression of a distinct viewpoint, the agency through which such viewpoints are expressed, the right and/or ability to express an opinion (through speech or other symbolic means), and/or an author's characteristic tone or style (Voice, n.d.). To emphasize the role of narrative style in constituting evaluative grounds and thus credibility, I define voice as *narrative style with efficacy*. Efficacy (or force) entails the capacity to achieve positive audience reception—the successful assertion of credibility. Enacted through narrative style, voice signals germane evaluative grounds and thus allows for tactical movement across different spheres of argumentation. A rhetor's objective is to establish credibility: an audience's perception of voice. Credibility is thus the aggregate of narrative style and a rhetor's ability to establish argumentative grounds. When successful, a rhetor attains voice.

Arguing for a nuanced understanding of voice (beyond either unqualified speaker agency, or a decentered "function of the linguistic") Eric King Watts defines the term as "a relational phenomenon occurring in discourse"—an event through which "we can characterize our commitments and sentiments toward our social spaces" (2001, pp. 191, 180, 192). Voice is here understood as an interactive process, one imbued with ethical and normative implications. Within this model, a rhetor's voice may be conceptualized as the interplay between the act of storytelling and an audience's varied attention and disposition.

In conceptualizing voice as an active practice (as opposed to a phenomenon or value), I build upon Nick Couldry's (2010) notion of voice as process. In contrast to common associations of voice with authenticity (e.g., an individual's "inner voice"), voice here is a strategic practice designed to achieve rhetorical effect. Couldry articulates two key models: voice as value (a commitment to the worth of democratic narrative practice), and voice as "the process of giving an account of one's life" (2010, p. 7). The former implies an egalitarian valuation of diverse viewpoints. At first glance Crichton seems to draw upon this ethos (his voice being but one that deserves an audience in the name of open inquiry). Soon clear, however, is Crichton's distaste for populism; he claims unique interpretive authority.

Building also upon Walter Fisher's work on narrative (1984, 1987), I envisage voice as the rhetorical process of attracting an audience and constituting their evaluative grounds. More than a vehicle for expression, voice entails the ongoing practice of symbolically forming ideas, social relations, common worlds, and feelings. In accenting narrative style, I emphasize a fairly traditional conception of voice (what Watts labels "speaker agency or authority") (p. 191). However, I believe that this remains a fundamentally important dimension. Following Fisher, I understand narrative as an active practice by which advocates set context, signal genre, and thereby constitute relevant grounds for evaluation. Crichton's speech and novel do this well, and decisively obscure the border between argumentation and aesthetics (Fisher, 1984, p. 2).

Within Fisher's model, the bases for narrative evaluations are " 'good reasons' which vary in form among communication situations, genres, and media" (1984, p. 7). "Good reasons," in turn, are determined by narrative probability ("what constitutes a coherent story"), and narrative fidelity ("whether the stories they experience ring true with the stories they know to be true in their lives") (p. 8). The crux of Crichton's argument is that elite environmentalists (including academics, politicians, and scientists) strive to corrupt and/or misrepresent science for personal gain. On this point, to the extent that Crichton builds upon strains of anti-intellectualism and distrust of politicians' motives, his story achieves fidelity with sympathetic audiences. The test of probability is less assured. As a science-fiction author whose views contradict the overwhelming evidence surrounding climate change, his story risks incoherence. However, I am ambivalent about Fisher's assertion that "the 'people' have a natural tendency to prefer the true and the just" (p. 9). In the context of discussions about complex climate science, the majority of citizens lack competence to interpret and evaluate existing data and thus must rely on others with assumed expertise (in other words, credibility). With respect to countries such as the United States, where the majority of citizens are scientifically illiterate (Duncan, 2007), Fisher's optimism seems misplaced. By no means qualified to interpret thousands of academic studies, Crichton nonetheless enacts such status and does so with force. Breaking stylistic conventions or decorum may lead to incoherence or parody. In this instance, by performing the role of disinterested public intellectual, a popular novelist circumvents the generic conventions that would traditionally limit his sphere of presumed expertise. A key element of Crichton's strategy is such chameleonic movement between argumentative spheres.

In 1982, Goodnight articulated the political significance of locating issues within a specific sphere of argument. For instance, poverty may be deemed a public issue only when "grounded in questions of public interest and responsibility" (p. 222). Similarly, when framed as an issue dependent upon expert assessment of scientific data, environmental concerns may transfer to the realm of technical expertise. Publication of *Silent Spring* initiated a reversal of this transition, as it translated a technical issue (DDT contamination) into the public sphere. A critical insight follows: "whereas scientists at least in theory should be able to create communities of inquirers without regard to the demands of the public, public leaders nevertheless provide parameters for scientific argument" (Goodnight, p. 222). Crichton speaks as such a "public leader." Through the tactical application of voice, he maneuvers between a range of rhetorical personae: celebrity author, dispassionate rationalist, maverick truth-seeker, public intellectual, government witness on climate science. In so doing, he purports to champion a set of compelling attributes: scientific integrity, the democratic principle of free inquiry, and common sense.

Robert Hariman proposes that "a text ultimately succeeds as an assertion of voice" (1995, p. 108). Without such success, without an audience, rhetorical practice fails to manifest as voice. To achieve voice is thus to acquire and enact power. Associating voice with subaltern and marginalized perspectives, much critical work tacitly defines the idea as political capacity (Fabj, 1993; Wertheimer, 1997). As argued by Bonnie Dow, "all social locations are not equal... our socially located voices have political implications" (1997, p. 243). Crichton's voice is definitively not "the sound of the dispossessed" (Watts, 2001, p. 191). While he purports to embody abstract ideals (reason, moderation, prudence), I interpret such moves as efforts to appropriate ethical high ground and cultivate rhetorical force.

"Environmentalism as Religion"

Crichton's 2003 speech to the Commonwealth Club of San Francisco addresses the North American environmental movement's increasing social prominence. Through an extended analogy, he draws a correlation between environmental and religious thought, dismisses both as dogmatic (the antithesis of an imagined "pure" scientific project), and thus asserts that environmental perspectives lack policy relevance. His core argument is that "the beliefs of a religion are not dependent on facts, but rather are matters of faith" (2003, p. 18). According to these broad standards, a vast swath of the human world is equally religious in epistemological orientation: in its ideas of human rights, democracy, capitalism, and cosmopolitanism (to name but a few). In this synecdoche, environmentalism as a whole is equated with an abstract hybrid of Deep Ecology and organizations at the militant fringe of the broader environmental movement. By evaporating the distinctions among a myriad of environmental groups ranging the spectrum from radical (Earth Liberation Front and Earth First!) to more conservative (the Nature Conservancy), and reducing the movement as a whole to one of its more radical versions, Crichton presents a choice between two opposing paths: Enlightenment (common sense and science) versus Dark Ages (religious fanaticism and feudalistic cabals). As a discerning outsider distanced from the corrupting influence of institutional "group think" (O'Keefe, 2006b, p. 1) Crichton epitomizes the former path. By contrast, environmental advocates are compromised by either extremist dogma or bureaucratic constraints.

Crichton makes only tacit reference to eco-terrorism. However, his equation of environmental thought with a relatively extreme political orientation[1] allows him to argue that "fundamentalists" such as contemporary American environmentalists are "dangerous because of [their] rigidity [and] imperviousness to other ideas" (2003, p. 19). Danger arises from such individuals' perceived severance from reality—a disconnect that threatens the well-being of others. Analogies between certain environmental campaigns and doomsday prophecies are easy to draw (e.g., "the end is nigh") and Crichton highlights this loose parallel for strategic gain. Associating environmentalists with "the nut on the sidewalk carrying the placard that predicts the end of the world" (2003, p. 18), he strives to command the "ideological legitimacy of populism" (McCarthy, 1998, p. 126). Mirroring tactics from the "wise use" movement, he frames his position as prudent, sensible, and judicious.

A core dimension of Crichton's attack on environmental thought proceeds through patronization: individuals may mean well, but they lack elementary understanding of the world. By contrast, Crichton situates himself as a sage voice of reason. Questioning environmentalism's presumed ideological commitments (specifically the idea that the world was idyllic prior to Western industrialization), he demands, "What was that Eden of the wonderful mythic past? Is it the time when infant mortality was 80%, when one woman in six died in childbirth?" (2003, p. 15). He quickly asserts that these are not idle fantasies; such beliefs are correlated with a dangerous and irrational worldview. At one register, environmentalists are simply inept.[2] At another, they verge on the criminal. Crichton's tone thus shifts toward a more aggressive attack on the ethical grounds of environmental thought: politically biased environmental organizations willingly lie and distort science with full knowledge that their actions will result in the deaths of innocents. Developing this narrative (and alluding to the banning of DDT and its implicit correlation with malarial deaths in Sub-Saharan Africa), he asserts: "We know from history that religions tend to kill people, and environmentalism has already killed somewhere between 10–30 million people since the 1970s" (2003, p. 19). Crucially, this line of argument exceeds Crichton's initial assertion that environmentalists simply don't "recognize that their way of thinking is just one of many other possible ways of thinking, which may be equally useful or good" (2003, p. 19) (a claim that highlights, by contrast, his own reasoned and accommodating stance). Rather, truth is under attack from elite environmentalists with political agendas. Threatening and extreme, such individuals demand challenge.

At this point, Crichton's arguments diverge. Is environmentalism just one of many possible paradigms? A criminally negligent religion guilty of killing tens of millions? Is the real problem any attempt at predicting or managing future outcomes? Or is the enemy irrationality that leads to bad (i.e., politicized or inaccurate) science? These multiple lines of attack strive to maximize Crichton's argumentative force. Articulating the (assumed) mythic origins of the modern environmental movement, Crichton foregrounds the charge of irrationality and draws an explicit parallel between environmentalism and religion:

> [E]nvironmentalism is in fact a perfect 21st century remapping of traditional Judeo-Christian beliefs and myths. There's an initial Eden, a paradise, a state of grace and unity with nature, there's a fall from grace into a state of pollution as a result of eating from the tree of knowledge, and as a result of our actions there is a judgment day coming for us all. We are all energy sinners, doomed to die, unless we seek salvation, which is now called sustainability.
>
> (2003, p. 15)

Beyond the acerbic tone, these words strive to dissociate environmentalism from empiricism and, by implicit extension, from rational thought. This tactic is precarious, as religious thought is often associated with moral standing. Anticipating this danger, Crichton thus commandeers the terrain of ethics: "I believe it is incumbent on us to conduct our lives in a way that takes into account all the consequences of our actions, including the consequences to other people, and the consequences to the environment" (2003, p. 14).

Such incursions into the realm of ethics, however, are a detour from his core strategy. Professing that "[w]e need to stop the mythic fantasies, and we need to stop the doomsday predictions" (2003, p. 19), Crichton affirms his allegiance to hard, unyielding science. Attacking environmental thought as corrupt, he asserts: "We must institute far more stringent requirements for what constitutes knowledge in the environmental realm.... [W]hat more and more [environmental] groups are doing is putting out lies, pure and simple. Falsehoods that they know to be false" (2003, pp. 19–20). According to this argument, prevailing scientific belief deems it both fashionable and profitable to advocate anthropogenically induced climate change. Those who dissent are ostracized.

Derived from this blueprint, two specific accusations are paramount. First, environmentalists are elitists. Second, their worldview is grounded in dogma. The first charge imagines a hostile division between environmental principles (which stifle human aspirations) and a commitment to human and economic activity. The second, correlated accusation equates environmental concern with religious fervor. In this narrative, environmentalism entails a covenant with self-propagating lies, wherein adherents reward and encourage each other's continued fabrications. This argument builds upon the concept of echo chambers (Sunstein, 2007) wherein self-selected publics tend toward a reinforcement, and amplification, of existing beliefs. I grant that several of Crichton's criticisms (e.g., fervent adherence to "the cause") are accurate with respect to the beliefs of some non-scientist activists. However, Crichton exploits this partial truth by conflating scientists with environmental zealotry. Critically important is the strategic redefinition of scientific consensus as orthodoxy. In contrast to Crichton's reasonable assertion that "deciding what constitutes responsible action is immensely difficult, and the consequences of our actions are often difficult to know in advance" (2003, p. 14), environmentalism is attacked as both fanatical and baseless. As observed by Myra MacDonald, "The meanings we attribute to words and images depend on cultural assumptions, and help, in turn, to perpetuate these" (2003, p. 9). By encouraging the public to equate scientific consensus with orthodoxy, Crichton mines a cultural aversion to the union of science with faith.

A scarcity of public knowledge about, and interest in, environmental science necessitates that risk communication professionals "construct a fair and understandable representation of both the relevant science and its meaning for the choice of policy options" (Powell & Leiss, 1997, pp. 222–223). In extending his claims of competence beyond fiction into the realm of science, Crichton circumvents the norms of the scientific community. Yet if he achieves a public role of trusted risk communication practitioner, these breaches, along with the vast body of literature he rebuts, are rendered invisible. Those who disagree are misguided at best, corrupt at worst. Crichton becomes the emblem of dispassionate rationality and forwards an eminently reasonable alternative: we shun convention and parochialism, and rely on apolitical facts. In the Oscar-winning documentary *An Inconvenient Truth*, Al Gore declares, "I don't really consider [climate change] a political issue, I consider it to be a moral issue" (Guggenheim, 2006). Such words, suggests Crichton, are anathema to the scientific project. Through hard work (he suggests), he has gained a privileged vantage point and deep understanding of nature and reality. This attitude is exemplified in his assertion "I have a fundamental answer" to "the greatest challenge facing mankind" (2003, p. 14).[3] Read with awareness that he contradicts the massive preponderance of evidence, Crichton's "fundamental" solution (an emphasis on rigorous science) rings hollow. There is a touch of audacious beauty in a fiction writer claiming greater understanding of climate dynamics than a consortium of hundreds of Ph.D. specialists, as Crichton associates himself with a long tradition of iconic truth-seeking heroes: Galileo (heliocentricism), Isaac Newton (gravitation), Albert Einstein (the general theory of relativity).

Appropriating the authority associated with scientific precision, Crichton employs the language of neutrality and prudence:

> [S]cience offers us the only way out of politics. And if we allow science to become politicized, then we are lost. We will enter the Internet version of the dark ages, an era of shifting fears and wild prejudices, transmitted to people who don't know any better.
>
> (2003, p. 20)

"Science" here operates as a clean abstraction (rather than, for example, as a specific methodology). Abstractions (e.g., liberty, wealth, progress) are difficult to counter. Such tactical imprecision is both common, and at times dangerous when the perspective in question swims against the current of existing scientific evidence. Significantly, calls for further research and open debate situate Crichton's arguments within a proactive, morally ascendant framework.

There is thus a perverse symmetry between his lament that "facts aren't necessary, because the tenets of environmentalism are all about belief" (2003, p. 15), and his declaration:

I can, with a lot of time, give you the factual basis for [my] views, and I can cite the appropriate journal articles.... But such references probably won't impact more than a handful of you, because the beliefs of a religion are not dependent on facts, but rather are matters of faith.

(2003, p. 18)

The argument could be paraphrased thus: trust me—if you don't, you're part of the problem. This tactic has long roots in American public address. In his immensely powerful precursor to the Declaration of Independence, Thomas Paine (1776) offered a parallel justification for the American Revolution: "Perhaps the sentiments contained in the following pages, are not yet sufficiently fashionable to procure them general favor; a long habit of not thinking a thing wrong, gives it a superficial appearance of being right". This earthy tone is more than a simple diversion; it is an attempt to draw upon the magnetism of common sense. The populist voice is perhaps Crichton's most effective, for through it he claims a potent discursive ground that straddles the worlds of scientific expertise and prudent judgment. Raising the specter of liberal conspiracy, he declares: "I can tell you some facts. I know you haven't read any of what I am about to tell you in the newspaper, because newspapers literally don't report them" (2003, p. 18). His sources for these "facts" are of the highest quality: "I can cite the appropriate journal articles not in whacko magazines, but in the most prestigious science journals, such as *Science* and *Nature*" (2003, p. 18).[4] His failure to do so, one must assume, is justified by the idea that his opponents (pseudo-religious fanatics) are immune to reason.

Defending level-headed wisdom, he condemns the misconceptions that cloud public judgment: "The truth is, almost nobody wants to experience real nature. What people want is to spend a week or two in a cabin in the woods, with screens on the windows" (2003, p. 16). The targets here are urban elitists who drive through rural areas "drinking Perrier and munching organic chips, staying occasionally in the bed-and-breakfast operations into which the homes of Westerners have been turned" (Pendley, 1995, p. 7). Differentiating his stance still further from that of arrogant liberal intellectuals, Crichton ridicules "an academic movement, during the latter 20th century, that claimed that cannibalism was a white man's invention to demonize the indigenous peoples. (Only academics could fight such a battle)" (2003, p. 16). It is no coincidence that one of *State of Fear's* antagonists is graphically killed by such cannibals.

Crichton, by contrast, identifies himself as down-to-earth and aware of the limitations of human knowledge. Evoking the twin spirits of humility and caution, he declares:

> We need to be humble, deeply humble, in the face of what we are trying to accomplish. We need to be trying various methods of accomplishing things. We need to be open-minded about assessing results of our efforts, and we need to be flexible about balancing needs.
>
> (2003, p. 19)

The populist voice has tactical benefits. Appeals for humility in light of the natural world's complexity and raw power justify evasions of scientific evidence, and warrant the postponement of difficult political decisions. Consolidating his rugged, populist persona, Crichton employs clear and accessible language: "Farmers know what they're talking about. City people don't. It's all fantasy" (2003, p. 16). Crichton here draws upon a classic American trope: an anti-elitist strain of Jeffersonian agrarianism which holds that the practical wisdom of the "salt of the earth" is superior to artificial academic knowledge. This move discloses a deep hypocrisy on Crichton's part—presumably the "real" scientists (who publish in "prestigious science journals, such as *Science* and *Nature*" (2003, p. 18)) are not uneducated farmers. The result is an erratic depiction of qualified environmental scientists, many of whom assert that environmental crises are real and deserving of attention and action. Yet should the argument's duplicity evade attention, it serves dual purposes. First, it differentiates Crichton from (assumed) politically compromised academics. More important, it increases his narrative's broad comprehensibility. It is difficult to address the complexities of environmental science without occasional reference to technical language. By shunning complex scientific concepts and focusing, rather, on the perceived frailties of environmentalism, Crichton directs his audience's attention toward a specific (negative) evaluation of the environmental movement and its objectives.

State of Fear

Part of the brilliance of Crichton's genre-switching occurs through the multifaceted positions from which he presents his critique of science. Such dexterity is exemplified in *State of Fear* wherein Crichton embodies three distinct personae: omniscient narrator, fictional characters, and (in the appendices) science advocate/critic.[5] The following spotlights the appendices wherein Crichton speaks openly as a scientific authority.

State of Fear's plot is concise: global warming is a hoax executed by conspiratorial environmentalists who employ false science to enhance their organizations' profiles. The central agency is a fictional environmental group (the National Environmental Resource Fund, or NERF) that plans to simultaneously trigger a set of natural disasters including a massive hurricane and a tsunami. These storms are intended to coincide with the group's global media conference, thereby promoting public fear of the (fictional) dangers posed by climate change. The central villain is Nicholas Drake, head of NERF. The hero is John Kenner, an M.I.T. professor/National Security Intelligence Agency superagent. Kenner specializes in countering the scientific lies propounded by mainstream news media. The pivotal figure is Peter Evans, a NERF attorney whose loyalty to the environmental movement slowly erodes throughout the novel (in parallel to his advancing knowledge).

At the novel's end, Kenner gives an impassioned monologue on how the environmental movement must be dismantled and reconfigured, and warns of the challenges ahead: "Nobody dares to solve the problems—because the solution might contradict your philosophy, and for most people clinging to beliefs is more important than succeeding" (2004, p. 566). Following this final chapter, Crichton takes the unusual step of shifting authorial persona and diverging from a traditional novelist's role by way of an "Author's Message." Therein he writes openly as a science advocate/critic, articulates his thoughts on environmental science, and presents himself as an authority on, and defender of, the scientific project. He asserts: "A novel such as *State of Fear*, in which so many divergent views are expressed, may lead the reader to wonder where, exactly, the author stands on these issues" (2004, p. 569). Countless novels embody "divergent views" through a story's course, yet literary works (especially pulp-fiction thrillers) are typically read as artistic creations. Millions of readers enjoy the works of Stieg Larsson or J. K. Rowling without much concern for the authors' personal thoughts about Swedish right-wing extremism or adolescent sorcery and witchcraft. Why might this novel be an exception, one that not only summons readers' curiosity about Crichton's "real world" beliefs, but also warrants the author's didactic intervention? I believe the answer involves Crichton's ambition to expand his domain of credibility beyond the realm of popular fiction.

Wallack, Dorfman, Jernigan, and Themba define advocacy as "a strategy for blending science and politics with a social justice value orientation to make the system work better, particularly for those with the least resources" (1993, p. 5). This definition mandates the melding of science with politics. Although Crichton candidly advocates for select social changes (e.g., revamping of the Environmental Protection Agency), he pretends to conceptually divorce these two realms of human activity. Responding to *State of Fear*'s infiltration into the political arena, many reviewers called this bluff. *Issues in Science and Technology* proclaimed:

> Rarely...have celebrities sought to use their fame as a platform to express themselves on the scientific aspects of controversial topics. And if they have, the public attention and impact have been negligible. With Michael Crichton, celebrity science has reached a new and disturbing level.
> (Miller, 2006, p. 95)

The *Pittsburgh Post-Gazette* called the book "a gripping, if sometimes tedious techno thriller that lumps global warming worrywarts in the derisive and divisive camp of baby seal-loving, tree-hugging, vegetable-munching pantywaists" and recommended it to readers whose "politics veer more to rabid than rational" (Walton, 2004). The *New York Times* was equally caustic. Addressing the novel's "annoying citations," Bruce Barcott (2005) stated: "To claim that it's a hoax is every novelist's right. To criticize the assumptions and research gaps in global warming theory is any scientist's prerogative. Citing real studies to support the idea of a hoax is ludicrous".

In the "Author's Message," Crichton asserts that because humanity can't predict the future, society need not take steps to reduce greenhouse gas emissions: "We can't 'assess' the future, nor can we 'predict' it. These are euphemisms. We can only guess. An informed guess is just a guess" (2004, p. 570). Contesting this logic, the Union of Concerned Scientists (UCS) (2008) responded, "the occurrence of large earthquakes is also very difficult to predict" and asked, "should we stop building earthquake-resistant buildings?". Contingency is central to life, whether the act in question involves conducting business transactions, commencing a military campaign, or simply stepping out of one's door. The UCS (2008) thus concluded, "Readers may understandably take away some misconceptions".

Other reviews were more charitable. For instance, *USA Today*, the highest circulating newspaper in the United States, lauded the "scientific data [Crichton] weaves through his stories." Carol Memmott (2004) asserted that the novel's central appeal is "Crichton's concern that the environmental movement has gotten off track". In such moments, when a pulp-fiction thriller masquerading as political manifesto is read as a serious scientific text, conventional genres are transgressed.

Indicative of the novel's blunt message is "Appendix I: Why Politicized Science is Dangerous." Drawing upon his strengths as storyteller, Crichton draws an analogy between contemporary climate science and the debunked, early 20th-century theory of eugenics:

> Once again, the measures being urged have little basis in fact or science. Once again, groups with other agendas are hiding behind a movement that appears high-minded. Once again, claims of moral superiority are used to justify extreme actions. Once again, the fact that some people are hurt is shrugged off because an abstract cause is said to be greater than any human consequences.
>
> (2004, p. 579)

These words are general. Depending upon one's sympathies, the passage could easily refer to a myriad of ideological positions. However, by creating the parallel between eugenics and climate science, Crichton directly associates the latter with artificiality and racist fascism. Taking the argument one step further, Crichton alludes to environmental justice: "most environmental 'principles'... have the effect of preserving the economic advantages of the West and thus constitute modern imperialism toward the developing world" (2004, p. 571). Although he provides no evidence to support this claim, the objective is clear. Environmentalists move beyond naiveté toward calculated depravity.

While acknowledging that he does not actively gather scientific data, Crichton distinguishes between the scientific practice of data accumulation, and the interpretive act of evaluating such data's meaning and implications. In respect to the latter task, he suggests that he is signally qualified to make authoritative judgments (indeed, the novel feigns scientific rigor by way of an extended bibliography). However, despite Crichton's assertion that "There is no evidence that my guess about the state of the world one hundred years from now is any worse or any better than anyone else's" (2004, p. 570), his claim to scientific expertise can be summed up by the passage, "I have been reading environmental texts for three years.... I have had an opportunity to look at a lot of data, and to consider many points of view" (2004, p. 569). This is a populist appeal to democratic competence: Crichton has gone to the source and thereby sidesteps the distortions of (politically compromised) bureaucrats. Yet to complete the logic, an audience must bestow sufficient credibility on Crichton for him to invalidate the existing body of peer-reviewed climate science literature. Although Crichton asserts that society should conduct research "in a humble, rational, and systematic way" (2004, p. 572), his position requires that an audience privileges the scientific judgment of a pulp-fiction author over the collective judgment of the National Academy of Sciences. Achieving this challenging task requires the tactical employment of voice.

Voice as rhetorical force

The scientific project has an ambiguous reputation in Western society. Central to modern culture, it is both championed as the key to a better life, and criticized in equal measure as an instrument of domination and reductive logic. Others have tracked the political and communicative dimensions of scientific thought (Kuhn, 1996; Goodnight, 2005). Commenting on the debate surrounding evolution versus intelligent design, John Lyne observes that argument is frequently less about empirical evidence than about "origins and mythic grounding, and about the very limits of science, fought out in the space of meanings and beliefs" (2005, p. 38). This pattern extends also to environmental arenas, wherein "battles are to be won and lost more on emotion than on a disinterested and objective scrutiny of scientific facts" (Gregory & Miller, 1998, p. 164). Indeed, the very idea of consensus, the fundamental argument in support of governmental climate-responsive regulatory action, is an explicitly political notion, more closely associated with democratic principles than with any ideal of universal and objective truth. Nonetheless, comprehensive understanding of scientific issues often requires a specialized ability to trace connections between broad global and historical patterns. An untrained observer is thus required to trust the recommendations of credible individuals. Crichton portrays himself as an exemplar of such persons—and he does so through voice fashioned to attract.

Rhetorical terrain expands as grounds of conviction contract. Persuasion is thus fundamental to debate over "controversial" environmental science. As a requisite dimension of civic deliberations, rhetorical practice may be employed toward multiple ends. On one key point, my thinking thus aligns with Crichton's stated views: language may be marshaled on behalf of ambitions which run counter to the collective interests of broader society, and one set of interests may be legitimized without appropriate transparency and without due public assessment of the best available evidence.

Inertia is a powerful vehicle. It is worth re-emphasizing that Crichton benefits from advocating for perpetuation of the status quo—his ultimate argument is that society should do nothing about (imagined) environmental exigencies. This proposition has two central thrusts. First, scientific knowledge is insufficient to justify actions that may engender substantial economic costs. Second, a range of flaws (naiveté, dogmatism, and/or political motivations) compromises the credibility of those who suggest otherwise. In these first two moves, having compressed Western environmentalism into a narrow bandwidth (loosely associated with the principles of Deep Ecology), Crichton swiftly discredits the entire movement. Notwithstanding his assertions to the contrary, these arguments are structured along religious lines. His premises must be accepted without evidence and on faith. The logic is Manichean; battle lines are clearly drawn and the outcome preordained. Beyond this standard tactic, a somewhat more interesting paradox infects his arguments: the claim to scientific purity, and tandem derision of the world's leading scientific organizations. The result is a strategic (rhetorically sophisticated, yet ethically corrupt) vacillation between opposing depictions of science as at times beneficial, at others deficient and unprincipled.

Yet while the coherence of Crichton's argument dissolves when situated within the broader context of prevailing scientific belief, there is risk that his views exert inordinate influence on public understanding of contemporary environmental issues. In debate over environmental policy, multiple voices interplay. Lacking technical understanding, the general public conventionally relies upon the counsel of experts. Translation of scientific data into policy outcomes thus requires an evaluation of source credibility. While there are rhetorical dimensions to both public-oriented and intra-community scientific communication (Kuhn, 1996; Ceccarelli, 2001), it is nonetheless vital to maintain robust standards of evaluation with respect to technical claims advanced in the public sphere (Whidden, 2012; Paliewicz, 2012). The bedrock for human judgment may be narrative-based "good reasons" (Fisher, 1984, p. 8), but some stories are better than others. By obfuscating the bases for appropriate judgment of scientific data, Crichton undermines the capacity for informed public dialogue regarding the prudent societal role of environmental perspectives.

The preceding has outlined a distinct (albeit preliminary) model for the rhetorical criticism of voice: an approach that articulates how evaluative grounds are constituted and/or transformed through the tactical operation of voice. Drawing a parallel between Couldry's (2010) idea of voice as process and Fisher's (1984) work on narrative, the essay traces each theory's connection to spheres of argument (Goodnight, 1982), and ultimately to rhetorical influence. Embodied through narrative style, voice signals genre and thereby constitutes germane evaluative grounds. The establishment of evaluative grounds, in turn, determines standards for credibility and may thereby circumvent traditional boundaries that delineate realms of competence. The result may be an escalation of rhetorical force. By envisaging voice as narrative style with efficacy, I aim to accentuate the continuous struggle over credibility. Voice is thus a performance that enables chameleonic movement between argumentative spheres. As rhetors orchestrate evaluative grounds to expand arcs of audience-conferred credibility, they engage in the voice-based exercise of communicative power.

Notes

1. Over the last several decades, U.S. newspapers have increasingly framed ecotage (particularly direct action sabotage by organizations such as Earth First!) as terrorism (Wagner, 2008).
2. This discursive maneuver echoes that of the George Marshall Institute (MI), a Washington, DC. think-tank whose tactics parallel Crichton's. Evoking the most recent American experience of endemic destitution, the MI asserts that just as the Great Depression "was caused by well meaning people who pushed wrong headed tariffs.... The consequences of [Al Gore's climate-related] policy preferences would be just as damaging" (O'Keefe, 2006a, p. 1). The core feature of this passage is the claim that such devastation was caused by "well meaning people." Such a move situates the MI within a sphere of ethical ascendancy. Implied therein is that the organization enjoys a privileged, objective view of the debate. Others may contribute with the sincere belief that their views are both correct and imperative. However, lacking the MI's aerial and historical perspective, such opinions are fated to replicate human ignorance and deficiency. Crichton's arguments follow a similar strategy: patronization, and attempted marginalization, of competing viewpoints as uneducated and misguided.
3. One may note this phrase's religious connotations—particularly in light of his previous doomsday prediction: "environmentalism has already killed somewhere between 10–30 million people" (Crichton, 2003, p. 19).
4. In a study of Exxon Mobil's public relations tactics, Sharon Livesey argues that, to counter compelling scientific evidence, that corporation has employed strategic selection of experts. In doing so, Exxon Mobil becomes both "judge and jury in [the] social debate" (2001, p. 129). State of Fear's extensive feigned-scientific bibliography exemplifies this maneuver. In "Environmentalism as religion," Crichton takes this strategy one step further. Rather than providing tactically selected proofs, he simply ignores all evidence and asserts "trust me."

5. A notable parallel may be drawn between *State of Fear* and Rachel Carson's *Silent Spring* in that the former starts with raw fiction but concludes with a pseudo-scientific bibliography (supplemented with two didactic "author messages"), whereas the latter starts with a parable before delving into more technical scientific material. *Silent Spring*'s 1962 publication was a foundational moment for the nascent Western environmental movement; it inspired a coalition of individuals concerned about the detrimental environmental effects of modern industrial technologies. As much poetics as science, the book fused the genres of environmental science and fiction (although toward quite different ends than those of Crichton 40 years later).

References

About Michael Crichton (2012). Michael Crichton: The official site. Retrieved 1 August 2013 from http://www.crichton-official.com/aboutmichaelcrichton-biography.html
Barcott, B. (2005, January 30). "State of Fear": Not so hot. *The New York Times*. Retrieved 1 August 2013 from http://query.nytimes.com/gst/fullpage.html?res=9903E2DA1038F933A05752C0A9639C8B63.
Ceccarelli, L. (2001). Rhetorical criticism and the rhetoric of science. *Western Journal of Communication, 65(3)*, 314–329.
Couldry, N. (2010). *Why Voice Matters: Culture and Politics after Neoliberalism*. London, UK: Sage.
Crichton, M. (2003, September 15). Environmentalism as religion. Retrieved 1 August 2013 from http://scienceandpublicpolicy.org/commentaries_essays/crichton_three_speeches.html
Crichton, M. (2004). *State of Fear*. New York, NY: HarperCollins.
Dow, B. J. (1997). Politicizing voice. *Western Journal of Communication, 61(2)*, 243–251.
Duncan, D. E. (2007, February 21). 216 Million Americans are scientifically illiterate (Part I). *MIT Technology Review*. Retrieved 1 August 2013 from http://www.technologyreview.com/view/407346/216-million-americans-are-scientifically-illiterate-part-i/
Fabj, V. (1993). Motherhood as political voice: The rhetoric of the mothers of Plaza de Mayo. *Communication Studies, 44(1)*, 1–18.
Fisher, W. (1984). Narration as a human communication paradigm: The case of moral public argument. *Communication Monographs, 51*, 1–22.
Fisher, W. (1987). *Human Communication as Narration: Toward a Philosophy of Reason, Value, and Action*. Columbia, SC: University of South Carolina Press.
Forest, S., & Feder, M. A. (2011). *Climate Change Education: Goals, Audiences, and Strategies*. Washington, DC: National Academies Press.
Goodnight, G. T. (1982). The personal, technical, and public spheres of argument: A speculative inquiry into the art of public deliberation. *Journal of the American Forensic Association, 18*, 214–227.
Goodnight, G. T. (1995). The firm, the park, and the university: Fear and trembling on the postmodern trail. *Quarterly Journal of Speech, 81*, 267–290.
Goodnight, G. T. (2005). Science and technology controversy: A rationale for inquiry. *Argumentation and Advocacy, 42(1)*, 26–29.

Gregory, J., & Miller, S. (1998). *Science in Public: Communication, Culture, and Credibility*. New York, NY: Plenum Trade.
Guggenheim, D. (Director) (2006). *An Inconvenient Truth*. Film, Lawrence Bender Productions.
Hariman, R. (1995). *Political Style: The Artistry of Power*. Chicago, IL: University of Chicago Press.
IPCC History (2012). IPCCFacts. Retrieved 1 August 2013 from http://www.ipccfacts.org/history.html
Janofsky, M. K. (2005, September 29). Michael Crichton, novelist, becomes senate witness. *The New York Times*. Retrieved 1 August 2013 from http://www.nytimes.com/2005/09/29/books/29cric.html
Kuhn, T. S. (1996). *The Structure of Scientific Revolutions* (3rd ed.). Chicago, IL: University of Chicago Press.
Leiserowitz, A., Maibach, E., Roser-Renouf, C., & Smith, N. (2011). Climate change in the American Mind: Public support for climate & energy policies in May 2011. Yale Project on Climate Change Communication. Retrieved 1 August 2013 from http://environment.yale.edu/climate/files/PolicySupportMay2011.pdf
Livesey, S. M. (2001). Global warming wars: Rhetorical and discourse analytic approaches to Exxon Mobil's corporate public discourse. *The Journal of Business Communication, 39(1)*, 117–148.
Lyne, J. (2005). Science controversy, common sense, and the third culture. *Argumentation and Advocacy, 42*, 38–42.
MacDonald, M. (2003). *Exploring Media Discourse*. London, UK: Hodder.
McCarthy, J. (1998). Environmentalism, wise use and the nature of accumulation in the rural West. In B. Braun & N. Castree (Eds.), *Remaking Reality: Nature at the Millennium* (pp. 125–148). New York, NY: Routledge.
Memmott, C. (2004, December 14). Crichton fans will embrace "Fear." *USA Today*. Retrieved 1 August 2013 from http://usatoday30.usatoday.com/life/books/reviews/2004-12-13-crichton_x.htm
Miller, A. (2006). Bad fiction, worse science. *Issues in Science and Technology, 22(2)*, 93–95.
O'Keefe, W. (2006a). Climate zealotry produces bad policy. *George C. Marshall Institute*. Retrieved 1 August 2013 from http://www.marshall.org/article.php?id=451.
O'Keefe, W. (2006b). Group think masquerading as consensus. *George C. Marshall Institute*. Retrieved 1 August 2013 from http://www.marshall.org/article.php?id=450.
Paine, T. (1776). Common sense. Retrieved 1 August 2013 from http://www.earlyamerica.com/earlyamerica/milestones/commonsense/text.html
Paliewicz, N. (2012). Global warming and the interaction between the public and technical spheres of argument: When standards for expertise really matter. *Argumentation and Advocacy, 48(4)*, 231–242.
Pendley, W. (1995). *War on the West: Government Tyranny on America's Great Frontier*. Washington, DC: Regnery Publishing.
Powell, D., & Leiss, W. (1997). *Mad Cows and Mother's Milk: The Perils of Poor Risk Communication*. Montreal, QC: McGill/Queen's University Press.
Rawlings, N. (2010). Top 10 failed celebrity political campaigns. *Time.com*. Retrieved 1 August 2013 from http://www.time.com/time/specials/packages/article/0,28804,2009170_2009172_2009179,00.html

Solomon, S. Qin, D., Manning, M., Chen, Z., Marquis, M., Averyl, K. B., Tignor, M., & Miller, H. L. (Eds.) (2007). *Contribution of Working Group I to the Fourth Assessment Report of the Intergovernmental Panel on Climate Change*. New York, NY: Cambridge University Press.
Sunstein, C. R. (2007). *Republic.com 2.0*. Princeton, NJ: Princeton University Press.
The Royal Society (2012). Climate change. Retrieved 1 August 2013 from http://royalsociety.org/policy/climate-change/
Time.com (2012). Top 10 airplane books. Retrieved 1 August 2013 from http://www.time.com/time/specials/packages/completelist/0,29569,1923223,00.html
Union of Concerned Scientists (2008, June 19). Crichton's thriller *State of Fear*: Separating fact from fiction. Retrieved 1 August 2013 from http://www.ucsusa.org/global_warming/science_and_impacts/global_warming_contrarians/crichton-thriller-state-of.html
U.S. Department of State (2009). Office of the special envoy for climate change. Retrieved 1 August 2013 from http://www.state.gov/s/climate/index.htm.
Voice (n.d.). In *Oxford English Dictionary*. Retrieved 1 August 2013 from www.oed.com.
Wagner, T. (2008). Reframing ecotage as ecoterrorism: News and the discourse of fear. *Environmental Communication: A Journal of Nature and Culture, 2(1)*, 25–39.
Wallack, L., Dorfman, L., Jernigan, D., & Themba, M. (1993). *Media Advocacy and Public Health: Power for Prevention*. Newbury Park: Sage.
Walton, A. (2004, December 26). Crichton techno-thriller explores danger of mixing politics, science. Retrieved 1 August 2013 from http://www.post-gazette.com/stories/ae/book-reviews/state-of-fear-by-michael-crichton-563693/
Watts, E. K. (2001). "Voice" and "voicelessness" in rhetorical studies. *Quarterly Journal of Speech, 87(2)*, 179–196.
Wertheimer, M. M. (1997). *Listening to their Voices: The Rhetorical Activities of Historical Women*. Columbia, SC: University of South Carolina Press.
Whidden, R. A. (2012). Maternal expertise, vaccination recommendations, and the complexity of argument spheres. *Argumentation and Advocacy, 48(4)*, 243–257.
World Bank (2012). Climate change. Retrieved 1 August 2013 from http://www.worldbank.org/en/topic/climatechange

3
Invoking the Ecological Indian: Rhetoric, Culture, and the Environment

Casey R. Schmitt

In his 1999 book, *The Ecological Indian: Myth and History*, Shepard Krech III dissects the popular modern image of Native American as environmental prophet, living in special harmony with land, plants, and animals. The idea that Indians, as a whole, are somehow more closely concerned with or aware of ecological stability and sustainability, he writes, is a construct built from outsider perspectives and European perceptions of noble savagery. Yet, in the modern era, many Native American activists and leaders have indeed invoked a language of ecological conscience and consciousness, while encouraging Indian action toward environmental conservation. Despite Krech's assertion that the stereotype, whether well-intended or not, is ultimately "dehumanizing" (1999, p. 26), it seems that the "Ecological Indian" is no longer merely a construct imposed from the outside, but rather an important rhetorical ethos/identity for Native Americans engaging in political discussions under a Euro-American hegemony.

From one perspective, Native Americans who speak of ecological consciousness and harmony risk perpetuating their own voicelessness in modern American publics. If we consider voice in Eric Watts' (2001) terms, as "a *happening* that is invigorated by a public awareness of the ethical and emotional concerns of discourse" (p. 185), and, at its core, "a phenomenon of public hearing" (p. 190), the Indian who is ecological seems to be always already spoken for by the Ecological Indian symbol-figure; that is, even while voiced earnestly, ecological Indian rhetoric readily reifies existing stereotypes, allowing those stereotypes and their corollary assumptions (and not the Indian rhetors themselves) to continue to speak for the Indian community at large. Participation in the dominant imaginary prevents the Native American voice from being heard clearly or precisely. Invoking the Ecological Indian risks enacting the kind of indirect exclusions described by Robert Asen (2002), functioning "tacitly through discursive norms and practices that prescribe particular ways of interacting in public forums" and compelling participants "to conform to established modes of discourse that effectively negate the perspectives and contributions of previously directly excluded individuals and groups" (p. 345). Acknowledgement of actual emotions, concerns, and argumentative points, in Watts' terms, can fall easily by the wayside.

Yet, from another perspective, invoking the Ecological Indian might be seen as a unique platform for promoting and amplifying Native American voices. Whereas Native Americans have struggled for agency, autonomy, and self-presentation when engaging in a variety of modern American public debates (see, for example, Morris & Wander, 1990; Lake, 1991; Miller, 1999; Dickinson, Ott, & Aoki, 2006; Rogers, 2006; Black, 2009; Brady, 2011; Salvador & Clarke, 2011; and Endres, 2012), the Ecological Indian represents a frame for deference to and authority for an otherwise marginalized and often dismissed minority population. While Native American voice may be silenced elsewhere, it is both expected and encouraged when spiritual reverence to and knowledge of the natural environment are involved—and through this frame, Indian anxieties, obligations, and community participation *are* exceptionally acknowledged.

This chapter will critically re-examine Krech's work as it applies to Native American ecological discourse while also exploring the simultaneous encouragement and discouragement of Native American voice that result from engagement with the Ecological Indian. It will couple Krech with similar arguments made by Calvin Martin (1978), as well as with the concerns articulated by Paul Nadasdy (1999; 2011) over the structural and institutional barriers to indigenous authority. Whereas Nadasdy highlights the unbalanced influence of Euro-American terms, conceptual frames, and organizational structures upon indigenous peoples, compelling them to meet hegemonic demands and expectations when engaging in ecological discourse, this chapter suggests that some Native Americans have found a means of reappropriating their own image from Euro-American culture, making the imposed role of Ecological Indian a source of empowerment and engagement in the public sphere. As Krech argues, the stereotype has indeed limited the available scope for Native American voices at times in the past and the present, but other leaders, speakers, and activists have knowingly embraced and reshaped the role, presenting Indian ecology as an alternative to Western perceptions and attitudes, and explicitly linking environmental concerns to social issues that Westerners have not historically considered as "environmental" matters, including housing, healthcare, social welfare, and economics. In other words, rhetorical invocation of the Ecological Indian identity has allowed Native activists to gain authority and influence in a system that is otherwise stacked against them and their communities.

The latter half of this chapter considers Anishinaabe activist Winona LaDuke as a case example, highlighting how, in a series of speeches given across the U.S. between 1999 and 2012, LaDuke indeed invokes the Ecological Indian in her rhetoric.[1] Presented before primarily non-Indian audiences, LaDuke's speeches reliably draw from both her own tribal experiences and more general, pan-Indian concepts of nature appreciation, prophecy, and folk tradition. My analysis uses close reading for repeated thematic and anecdotal elements. The chapter concludes with a discussion of the implications of the double bind faced by the Indian who is ecological when speaking publicly and striving to be heard as an autonomous voice with original arguments of social consequence. The example of LaDuke, I argue, demonstrates that at least some Native Americans have managed to adopt what Krech might call a "dehumanizing" voice and to humanize it by making it their own; have taken the widely held assumptions of the public at large and bent them to promote their own agendas, peoples, and welfare.

Krech's Ecological Indian

At the heart of his argument, Krech seeks to reestablish Native North Americans neither as ecological sinners nor saints, but as people like any others. Native American communities, he writes, have indeed worked to preserve their environmental resources, have indeed developed a unique relationship with the biophysical systems of the North American continent, and have indeed suffered disproportionately from resource over-exploitation, pollution, and degradation in recent years, but they have also polluted and altered the land themselves throughout history, often with no regard for sustainability or conservation in Western terms. Native American cultures and communities, he writes, are ultimately as diverse as any others, and yet the blanket attribution of environmental ethics and insight persists.

Krech traces a series of popular media depictions of Indians as gentle conservators of a pristine environment, spoiled by European and

Euro-American actions (1999, p. 16). From Columbus to Rousseau to James Fenimore Cooper, he writes, descriptions and depictions of Native American peoples have consistently (and not coincidentally) served to provide Westerners with an Other figure against which to measure and assess their own behaviors and policies, a seemingly serendipitous mirror for socio-political critique. The commercial popularity of the Cooperesque noble savage, he writes, has waxed and waned alongside calls for resource policy reform (1999, p. 19). During the 1960s and 70s, especially, coinciding with a boom in global environmental activism, the Ecological Indian became an icon of "environmental and antitechnocratic causes" (1999, p. 20). As Calvin Martin (1978) puts it, the Indian "was introduced to the American public as the great high priest of the Ecology Cult" (p. 157), feeling deep sympathy for all living things and brimming over with "ecological prescience and wisdom" (Krech, 1999, p. 21).

This concept of the Indian, however, served the purposes of Europeans and Euro-Americans more than it did the Indians themselves. It allowed White Americans to acknowledge Native American peoples with reverence and, in effect, to negate what Dickinson, Ott, and Aoki have called "the social guilt" that might result from outright exclusion of Indian peoples from public discussion (2006, p. 29). By understanding the Indian as a humble ecologist and silent defender of an ever-retreating wilderness, dominant discourses assuredly acknowledged the Indian's existence but also circumscribed the nature of that existence in the public imaginary, removing from public discussion a frame for addressing non-ecological Indians in cities and modern towns, struggling with post-colonial hardships like obesity, unemployment, and incarceration. As Dickinson, Ott, and Aoki note, "Reverence exercises a double articulation, evoking both a profound sense of respect and a distanced, observational gaze" (p. 28); and, as Michelle Brady (2011) has argued with respect to museum exhibits on Native American peoples, that observational gaze can in fact silence or distort the very voices it purports to amplify. Films and novels of the later 20th century—especially those designed for children—repeatedly emphasized an almost superhuman Native American reverence for the land.[2] And at the governmental level, even Secretary of the Interior Stewart L. Udall declared Indians to be "in truth, the pioneer ecologists of this country" (Martin, 1978, p. 160).

Krech and Martin, among others, have made great efforts to demonstrate that such an understanding of Native American culture misreads history. Each presents abundant evidence to show that Native Americans have historically recrafted and damaged their environments as much as any other population. In recent years, too, many tribes have elected to sell rights to land and resources, and to place other concerns above biophysical sustainability. Applied universally to peoples from all tribes, the Ecological Indian disallows such diversity.

Still, this is not, perhaps, the most worrying aspect of the Ecological Indian's ascent. Rather, the stereotype does a fair deal of symbolic violence as well, by seizing the power of self-definition from millions of Native American people. The White image of the Indian defines the character of the Native American for national and global audiences, minifying his or her role to that of ecological symbol. It limits opportunities for Native American social engagement. Moreover, Krech argues, the Euro-American projection of Indian identity has become a self-image for many Native Americans. Many, he writes, "have taken on the Noble Indian/Ecological Indian stereotype, embedding it in their self-fashioning, just as other indigenous people around the world have done with similar primordial ecological and conservationist stereotypes" (p. 27).

Asen's warnings about the public imaginary's capacity to disadvantage and silence are here again worth consideration. As Morris and Wander (1990), Lake (1991), Black (2009), and others have all demonstrated, post-colonial Native American rhetorical efforts have consistently pushed against Euro-American constraints and toward decolonization, with varying degrees of success, but their very participation in national discourse—even in efforts of resistance—limited avenues for action and often constrained identities to the expected roles of nature prophet or noble savage. While Black notes that 19th-century efforts "worked by participating in the dominant discourse, and challenging the frameworks that the U.S. government set as foundational" (2009, p. 80), Krech's description of the Ecological Indian's popularity in the 20th century suggests that Native American speakers are increasingly constrained by prescribed roles. His argument mirrors Asen's claim that wider publics may "attempt to recuperate dissident appeals by creating an institutionally sanctioned and politically innocuous space for the expressions of opposition" (2002, p. 347). For Native Americans of all backgrounds, that space is the ecological.

All of this also echoes Richard A. Rogers' work, which demonstrates that Native Americans, as individuals and as groups, sometimes participate in "their own subordination due to powerful influences over what is 'taken for granted' and naturalized" (2006, p. 483). Even in acts of resisting Western expectations, rules, and values, the speaker who plays the role of Ecological Indian reifies the existing system, he might argue, because the system expects the Indian to be ecological. Rogers writes, "The performance of resistance using the imposed culture of the dominant indicates the presence of agency but not necessarily an agency grounded in the a priori subject of liberal individualism" (2006, p. 486). Krech's *Ecological Indian*, in this respect, becomes a treatise on voice, following Watts' assertion that such projects are "increasingly concerned with confronting, deconstructing, and interrogating a dominant language system that denies difference and, thus, mutes 'voice'" (2001, p. 183).

The implications of Krech's findings for social justice and indigenous rights thus seem, at first glance, to be somewhat dire. Still, with the remainder of this chapter, I would like to complicate the existing scholarship. Are we to dismiss wholesale all Native American ecological appeals as self-silencing and self-inhibiting perpetuation of a Euro-American ideal? Are we to say that all Native American speakers who would speak from a nature-appreciating perspective are always already dooming their own cause? Certainly, Krech and Martin make a convincing point when they argue that a widening of perspectives will benefit both indigenous peoples and their continued relationship with the land. Yet, to presume that Native American adoption of Western social roles is merely an extension of Western imposition overlooks any agency on the individual Indian or tribal community's part. Despite documentation of historical variation in ecological traditions, many Native Americans—like Winona LaDuke—have indeed embraced the modern environmental movement. For some, speaking especially of nature and environmental spirituality is a conscious rhetorical choice. For many, the welfare of the natural environment and the land are indeed primary social concerns, inextricably linked to local employment, healthcare, and religious freedom. In the following sections, I argue that within a Euro-American system that already constrains Native opportunities for engagement, invoking and enacting the Ecological Indian can in some cases be a strategically savvy move—and while Krech's initial arguments hold great merit, the issue of the Ecological Indian in modern environmental discourse is a bit more complicated than it first seems.

Worldviews, TEK, and politics

From the mid-1800s through the early parts of the 20th century, early environmentalists—including George Perkins Marsh, Aldo Leopold, and John Muir—thought of nature as an equilibrial whole, with harmonies disrupted and exhausted by human interaction. In recent decades, however, ecologists have recognized the failures of the equilibrium model and turned to gain insight from the traditional environmental knowledge (TEK) of indigenous and local peoples, accounting for chaos and dynamism in biophysical systems.[3] TEK represents a more holistic, localized, and common sense-driven approach to resource management. In communications scholarship, we can see a similar turn in Donal Carbaugh's advocacy for "listening" to the environment in a manner inspired by the Blackfeet (1999), in Salvador and Clarke's Weyekin Principle inspired by the Nez Perce (2011), and Danielle Endres' (2012) attention to Shoshone and Paiute value systems when discussing Yucca Mountain. While critics like Arun Agrawal (1995), for instance, have certainly problematized the distinction between scientific and indigenous knowledges,[4] this new shift in ecological focus has set the Native American in a new, potentially influential, policy-guiding role. It has provided a way for the Indian voice to be acknowledged, valued, and heard.

Whether imposed from the outside or not, the image of Native American as environmental authority has endowed the Native American community with an unusual degree of political capital. The Native American community has, over time and through various forms of representation, accrued an "environmental identity"[5] recognized both within the community itself and on the greater national and international level. Thus, while the Ecological Indian seemingly constrains the rhetorical avenues available to Native American leaders and activists, it does not wholly preclude their involvement and efficacy in American discourse entirely, as a more dire reading of Krech might suggest. The potential for Native American voice—for empowerment of the oppressed (Watts, 2001, pp. 181–182) and for increased civic participation (Brady, 2011, p. 204)—remains in the very same stereotypical structure that limits it. That is, while Native American concerns and opinions are for the most part muffled or dismissed in larger national discourses, large groups of both Native Americans and Euro-Americans currently consider TEK to be legitimate knowledge, recognizing in the Native American ecological speaker an opportunity to illuminate heretofore undiscussed strategies and solutions.

Of course, some inequities and barriers remain. Martin explains, for instance, that while the Ecological Indian is a construct, many Native American conceptions of the environment have still represented a "profoundly different cosmic vision when it came to interpreting Nature—a vision Western man would never adjust to" (1978, p. 188). Paul Nadasdy (1999) takes this point a bit further, arguing that, in fact, publicly stated efforts to incorporate traditional knowledge are in practice a mere power ploy, aimed at expanding Western authority over minority cultures and populations. Invitations by ecologists and policymakers to Native representatives, he claims, reinforce existing power relationships, as they compel Indians to work within Western biases and forms, to adopt Western terms that may not directly reflect indigenous understandings of the land, and to self-censor their ideas by catering to primarily White, Western audiences (p. 14). These representatives are forced to justify their non-Western worldviews within a Western framework, and this is inherently problematic. Even if dominant cultural appeals for the testimony of Native American ecologists are sincere (and Nadasdy implies a large portion of such highly political and visible appeals are not), they are heard and weighed by audiences who hold a thoroughly Western, non-Indian worldview. Nadasdy warns that the Indian activist who takes on the role of Ecological Indian and who appears to gain credibility in Western political terms is in fact often misled under an "illusion of agreement"—a system of jargon and simplification that emphasizes shared interests while obscuring existing, potentially irreparable schisms (pp. 136–148). The illusion of co-management, he writes, ensures that even when disagreements surface, the sentiment of common understanding and, thus, the existing structure of relations between Westerners and Native Americans, will remain. The Native American activist may speak up about environmental issues, he writes, but this counsel is heeded selectively, remains marginalized, and will never in fact exercise full authority over any land management decisions.

United States conservation policies have, in fact, at times directly harmed Native American culture, communities, and well-being.[6] From this perspective, endowing Native American peoples with the Ecological Indian persona can, indeed, enact a kind of environmental racism, marginalizing their role in socio-political discourse to mere affirmation of Western ideas of environmental conservation. When breaking from Western conceptions of conservation and ecology, Indian representatives have been condemned, attacked for being non-Indian, and held to ecological standards rarely expected of any White individual.[7] It is the exact kind of danger Rogers (2006) warns against in speaking of cultural dominance and that Morris and Wander noted when discussing how Indian voices are often only valued when the speakers "talk and act" like "real Indians" (1990, p. 165). The "valuing" of Native American viewpoints seems to correlate suspiciously with when those viewpoints promote already popular Euro-American policies with desirable social and economic outcomes for the dominant population.

Moreover, Nadasdy warns, as younger generations of Native activists seek to bridge the culture gap by seeking Western education, they are in effect assimilated and lose touch with the TEK they may aim to preserve in the first place (1999, p. 13). It's a kind of paradox, it seems, for the Native American who would seek to promote tribal concepts and interests without Euro-American constraints or influence.

Yet despite all of the barriers, several individuals and communities have found success when seeking to engage the public through a nature-reverent frame (Grossman, 2005). Despite the many ways in which we can see the Ecological Indian stereotype constraining the available means of Native American socio-political engagement, many of today's most prominent and influential Native American leaders refuse to censor their Indian-ness by speaking in wholly Western terms, dressing in wholly Western styles, or foregoing discussion of the links between people, religion, and environment that undergird many Native American worldviews. Indeed, rallying around the authority given them as Ecological Indians, Native American speakers and leaders have found a forum within the Western system where they can, indeed, be outspoken, and where they can, indeed, promote Native concerns on a national level.

We are left with the question of how, in spite of Krech and Nadasdy's warnings, this might be so. While a number of prominent communications scholars have examined patterns and specific instances of Native American rhetoric, much of this work has focused, like Krech's own argument, on the ways in which existing discursive systems distort Indian voices. Ono and Buescher (2001), as well as Rogers (2007), have examined how Native American identity in America has been commodified—simplified and recast "within a western, capitalist frame" (Ono & Buescher, 2001, p. 25). The most prominent representations of Native Americans in American society continue to be produced by non-Indians, and result in an "*abstraction*" of the Indian, "embodying notions of barbarism, nobility, stoicism, the inevitability of disappearance, harmonious spirituality, environmental stewardship, and other shifting and contradictory themes driven by Euro-American cultural dynamics" (Rogers, 2007, p. 236). Other scholars have documented Native American resistance to stereotyping and other means by which Euro-Americans have "spoken for them," even under purportedly "honorable motives" (Rogers, 2006, p. 488). In the early 1990s, Morris and Wander turned their attention toward rhetorical efforts of Native American individuals and communities, resisting and overcoming the "historic bloc" which had marginalized their voice in American society

(1990, p. 165). The result, they noted, was however also a threat to tribal identity, as a growing pan-Indian movement celebrated generalizations and similarities between tribes rather than specific tribal distinctions (pp. 166–167). It has only been in recent years—since, for instance, Randall Lake (1991) called attention to differences in Native American rhetorical styles—that scholars have pushed for a reconsideration of Native American rhetoric from Native American perspectives.

The modern Native American speaker's continued invocation and enactment of the Ecological Indian ethos most definitely grows out of a complex and violent colonial history, but it must first and foremost be heard and analyzed on its own individual terms.

The Ecological Indian in action: Winona LaDuke and the White Earth Reservation

Krech mentions Winona LaDuke, Anishinaabe activist, only at the very end of his 1999 book (p. 228). Rather than call special attention to her methods or successes, he places her within a long line of Native American writers—including Charles Eastman, M. Scott Momaday, and Vine Deloria, Jr.—who have emphasized indigenous links to nature and special ecological skills. The prescribed role of ecological authority allows each a freedom of expression where nature is involved, and many such Native American writers and leaders have embraced a rhetoric of general reverence for life and harmony with nature. For many, using a voice that emphasizes care and concern for the natural environment is not an unnatural or resisted means of expression; rather, while it may be influenced and encouraged by historical and socio-political constraints, it is comfortable and more favorable than legalistic or bureaucratic alternatives. That is, for the individual or group that earnestly does care for the environment, it is difficult to say where and when speaking from an environmental perspective grows from internal motivation and where and when it is an imposed, expected role. While Krech calls the stereotype "dehumanizing," it clearly does ring true for certain individuals and communities. Geographic location and historic circumstances have indeed allowed certain indigenous groups to develop special knowledge of and reverence for the land, and the general acceptance of Indian as conservationist has encouraged these groups to be vocal. The influence of accepted/prescribed social roles may at times seem, as Krech seems to feel, coercive, but it also opens a line of dialogue for historically marginalized populations when specific environmental arguments and cases arise.

Winona LaDuke represents one of a new group of Native American public figures, like Gail Small or Sarah James, who have parlayed expectations of Indian ecology into successful social movements by combining knowledge of Western legal systems with Native American tradition, and by using this hybridity of perspectives to write specifically on particular cases and causes. Harvard educated and politically active at the local, national, and international level, LaDuke has promoted infrastructural and green development on Minnesota's White Earth Reservation; worked to increase harvest yields on reservations across the American Midwest; directed both the Honor the Earth and White Earth Land Recovery Project (WELRP) organizations; helped to support successful Native anti-mining campaigns in Wisconsin, Minnesota, and Montana; and accrued enough prominence to twice run for United States Vice President on the Green Party ticket.

The majority of LaDuke's efforts and successes have focused upon the well-being of those living on her home reservation. The Anishinaabeg have long fought for resource sovereignty in the Midwest. Through allotment acts and broken treaties, they consistently lost land through the 19th and 20th centuries. When the states of Wisconsin and Minnesota applied uniform conservation laws throughout their territory, treating reservation and non-reservation land as homogenous political space, Anishinaabe access to key resources—and, particularly, to wild rice, spiritually and culturally significant to the people—was greatly restricted.[8] Legal challenges met with little success. Today, the largest landholders on the reservation are still federal, state, and county governments, and both mining and timber industries regularly campaign for access to rich reservation resources. Beyond these concerns, however, those on the reservation have grappled with other social exigencies: 50% unemployment, substandard and overcrowded housing, and an arrest rate seven times that of non-Indians in the area (Anonymous, 1999).

LaDuke, in her public rhetoric, links environmental concerns to these other social issues, allowing her to address the many needs of her community through a focus on ecological ethos. In recent years, through this strategy, she has helped to boost employment on the reservation, build schools, construct wind turbines, and install solar panels for energy on White Earth land (LaDuke, 2011d). Meanwhile, she has also promoted an endless number of specifically ecological causes, encouraging cultivation of nearly extinguished varieties of squash and corn, repopulating California rivers with salmon, and rallying protest movements against water pollution, strip mining, and nuclear waste.

Her linkage of ecological concerns to other Native American sociopolitical issues is particularly worth noting when assessing her "voice." LaDuke indeed performs the culturally sanctioned and expected role of spiritually driven Native American ecologist, but in performing the role she resists speaking *only* of spirituality, Native America, and ecology. She is not merely playing the cardboard-cutout part of the Ecological Indian to appease White appeals, as Nadasdy might have feared; rather she is using the Ecological Indian voice to assert agency in discussing a wide variety of concerns, declaring that Native American identity and ecology are indeed linked to broader questions of American and international citizenship. In effect, through this linkage, she claims the authority already given to her to speak on nature and ecology as simultaneous authority to speak on other socio-political topics on which Native voices have generally been, at best, sporadically considered. This at times enthymematic linkage and interweaving between the ecological and the social, economic, and political permeates all of her efforts.[9]

A close reading of LaDuke's public speeches over the past 14 years demonstrates how her platform draws from the authority of the Ecological Indian but also ties ecological concerns to social and political policy. She cites historical data on allotment regularly, even when speaking on water rights or wild rice (LaDuke 2004; 2011a; 2011b; 2011d). She presents flaws in the legal system and social structure as the root causes of ecological degradation, and, in effect, parlays her authority as Ecological Indian into a platform for greater social critique. In many of her speeches, she calls direct attention to her Ivy League training as an economist (2008; 2011a; 2011b) and utilizes scientific charts and data (2011a; 2011b) fluidly alongside Anishinaabe anecdotes and philosophies. While always spending the larger portion of her speeches discussing what Westerners might call ecological or environmental topics (land stewardship, agriculture, clean energy, and sustainable practice, for instance), she also explicitly addresses what dominant American culture might call "non-environmental" topics, including job growth, health and obesity, women's rights, demilitarization, religious freedom, energy profit and ownership, food sovereignty, and more (2011a; 2012), weaving seamlessly between concerns by tying them all together through their dependence on a sustainable and healthy environment. In a 2011 speech on energy policy and climate change in Seattle, Washington, in fact, she explicitly addressed each one of these seemingly varied ideas (2011a). The way LaDuke frames the discussion puts all of these concerns under the jurisdiction of the nature-minded Native American. For LaDuke, environmental justice and anti-colonialism are linked. In fact, she claims, despite a history of exclusion, Indians' engagement with socio-political topics is a predictable result of their collective history. In discussing resource management, she explains, "our community is also one that has been politicized by the whole process" and emphasizes that the lack of Anishinaabe success in reclaiming tribal land is largely "because we're kind of in the court of the thief. The U.S. Government was legally responsible for our state as Indian people" (Anonymous, 1999, p. 20). Though she speaks almost exclusively on "ecological" subjects, her greater social critique is less than subtle.

Yet, through all of her confident, assertive challenge of the status quo, and for all of her well cited understanding of Western legal policies, LaDuke also repeatedly resembles Krech's supposedly "dehumanizing" Ecological Indian. Her speeches trumpet an Indian identity that is linked to the land and the natural environment. This link is made clear from the opening moments of each public talk, as she begins every single speech with a traditional-style Anishinaabe greeting and thanks to her audience, spoken in Ojibwe and identifying herself with the specific land she is from—the White Earth Reservation in northern Minnesota. She can safely presume that her words in Ojibwe are inaccessible to Anglophonic audiences, and she always follows the greeting up with a rough translation, but the disjunct created by the opening words call immediate attention to a distinction—a difference—between LaDuke and her audience, immediately marking her as Indian (1999; 2008; 2011a; 2011b; 2011c; 2011d). In fact, the Ojibwe greeting marks her as different precisely because she *is* Indian.

Following this formulaic introduction, LaDuke regularly peppers her speech with other Ojibwe words and phrases, especially in recounting the Ojibwe names of the months (or moons) of the year in White Earth tradition (2008; 2011a; 2011b; 2011d; 2012). She makes a point of naming each moon distinctly, recounting the entire yearly cycle, and pointing out that each moon (like "maple syruping moon" "strawberry moon," or "wild rice moon") is named for its relationship to land, weather, and place, and not for Roman emperors. This point, of course, elicits chuckles from most audiences, but LaDuke uses it to again stress the nature-oriented worldview that distinguishes Native Americans and Native American thought from empire-driven Western perspectives.

In much of her discourse, LaDuke calls attention to her Ecological Indian-ness, filling talks on energy policy (2011a) or climate change (2008) with seemingly non sequitur references to broad, pan-Indian ecology and spirituality. When addressing specific cases, like the need to protect wild rice rights along Lake Superior (LaDuke, 2004; 2011d), she refers generally to Native ecological efforts across the country, from Montana to New Mexico to Alaska, as a whole. When speaking on Anishinaabe efforts toward sustainable agriculture in Wisconsin and Minnesota, she references other groups' ecological efforts in California, South Dakota, and the South Pacific, eliding the varied causes as one general indigenous environmental front (2011d). Much of her discourse draws from and emphasizes a generalized Native American spirituality as well. She speaks frequently of "the Creator," of ancestors, of animals as humans' relatives, and of the pathway to the "next world" (2011a; 2011b; 2011c). She refers to ecological principles as the "Great Law" (2008) or "Creator's Law" (2008; 2011b) and urges listeners to learn "how to walk the path that the Creator gave [each of them]" (1999). She calls environmental activism a "spiritual opportunity" (2011c) and remarks on "what a great spiritual opportunity it is to be the people that can be in that moment where you can do something great" (2011b). Stressing cyclical patterns in nature (including the hydrological cycle, lunar cycle, tidal patterns, etc.), she celebrates cyclical Native American worldviews and philosophies as a means to natural solutions, urging listeners to consider all living and non-living things as "related" (2008; 2011a; 2012) and to think about the "seventh generation" from now when making choices, just as traditional Native American maxims suggest that they do (2011a).

In taking this generalized, spiritually driven approach, LaDuke again risks playing the role of the nature mystic and thus being relegated to the sidelines of so-called "serious" ecological policy discussion, yet she explicitly—even reflexively—draws from this spiritually based appeal. In a 1999 speech in Seattle, Washington, for instance, she openly decried the ways in which Native American prophecy and religion are often dismissed as "folkloric," and protested that "we," the Native American people, hold such spiritually and traditionally held knowledges in high regard (1999). Relating stories about the Great Flood and Muskrat helping to create Turtle Island (2008), tales of White Buffalo Calf Woman's promised return bringing on an age of harmony (1999), or the myth of how Hawaiians developed taro (2011a), she says, is worth doing when speaking on Native American ecological and societal efforts because such stories "actually guide some of the work that a lot of our communities do" (1999). To not share these aspects of her community and culture, she seems to say, would be to misrepresent their worldview, to ignore some of their most important concerns, and to censor her own public voice.

Thus, when speaking on social or scientific topics, she draws heavily from poetic style and religion. When asked explicitly, for instance, "What is the link between the loss of biodiversity and the health of Native American communities?" she responds not with Western-style sociological, anthropological, or geographic models, but reverts to a description of Indian spirituality and tradition, saying:

> The Creator gave the Anishinaabeg people an immensely biodiverse forest. And he said, "Within this forest you will find all of your medicines. All the things you need to make your houses. All the foods you will need to sustain your families. The materials for all the baskets and other objects of amazing beauty that you can make. You can fashion all of those things from this land, upon which I'm putting you. Your job, though, is to take care of that which I gave you...." That is in essence our teaching.
> (Anonymous, 1999, p. 22)

The tone fits that of Krech's Ecological Indian, yet the agency and assertion comes not from the White interviewer but from LaDuke herself. She refers to ceremonies and songs when asked about land management policies. She says the rise of American agriculture is "not only about an economic and social transformation, it's also about a spiritual transformation" (Anonymous, 1999). She refers to reservation land as a sacred site, likening it to the Wailing Wall (Anonymous, 1999). She speaks of cultural values, and stresses the need for a cyclical model over a linear one, "because that's the natural world" (Anonymous, 1999).

LaDuke embraces her Indian identity before diverse crowds. She uses abundant Native American imagery when speaking, including paintings of Indian men in full regalia (2011a; 2011b), men in skins rowing a canoe (2011a; 2012), and stylized animals (2011a; 2011b; 2011d), while discussing diverse and modern topics. She calls special attention to herself as a "traditional practitioner" standing against the "trivializing" of Indian people and perspectives (1999). While playing the part of the Ecological Indian, she explicitly challenges a status quo in which Native Americans who would draw heavily and primarily from Native American tradition are not considered "qualified" to speak on matters of broad social concern (1999; 2011b). Native American opinion is "not held in the same regard" as Western opinion, she says, but, rather, "it is quite frankly held in lower regard" (1999), relegated to fringe discussions and anthropology wings instead of humanities, ecology, and political science departments (1999; 2008; 2011d).

Speaking in late 2011 at the University of Wisconsin-Madison (at an event I myself attended), LaDuke presented the image of an Indian man in regalia sitting before wind turbines as a jumping-off point, then used Anishinaabe prophecy as an organizing theme. Both were common tropes in LaDuke's public speeches that year, with the prophecy of the "People of the Seventh Fire" having played a prominent role in her rhetoric since the mid-1990s (1999). Speaking both humorously and earnestly of a fated choice between the well worn but scorched path and the less well worn but green route, she suggested that we are living now in the prophesied time of choosing before advocating against strip mining in Northern Wisconsin and in favor of food sovereignty in localized areas. Native American traditions and immediately pertinent political concerns were melded before the audience.

LaDuke's organization websites, too, draw heavily upon the images of Indians in some sort of special, spiritual connection to the land. The Honor the Earth site, for example, at the writing of this chapter, featured a downloadable copy of the organization's annual report, which begins by invoking Mother Earth at length, as a community with a "common destiny," gratefully "acknowledging that Mother Earth is the source of life, nourishment and learning and provides everything we need to live well; recognizing that all forms of exploitation, abuse and contamination have caused great destruction, degradation and disruption of Mother Earth, putting life as we know it today at risk" (Honor the Earth, 2011).

Discourse bordering on the mystical and romantic—talk of Mother Earth and animals as relatives for instance—rarely finds its way into contemporary policy discussion at the level of policymakers themselves, but as Native American speakers are often encouraged or expected to speak in such terms (Morris & Wander, 1990), leaders and activists like LaDuke have found uniquely Indian ways of asserting a voice in public discussion. By playing into the Western-imposed role of reverent nature worshipper, LaDuke puts herself at risk of dismissal, as Nadasdy warns— and yet, her initiatives grow in size and success each year. This unique turn in the Ecological Indian's political history draws from LaDuke's ability to invoke and enact what is expected of her, while also demonstrating what is not: a thorough literacy in Western science and legal issues. She creates a discursive hybrid of TEK and Western science not by incorporating one into the other, but by maintaining dual literacy and using the deference afforded one to demonstrate competence and drive inquiry in the other.[10] By merging discussion of environment with discussion of jobs and food, she is highlighting not only that marginalized communities are disproportionately affected by environmental degradation but also that we cannot separate the environmental from the social. In effect, LaDuke, as an Indian who is ecological, resists and redefines the Ecological Indian stereotype.

Conclusion: The Indian who is ecological and the Ecological Indian

I would like to address and dispel some potential misreadings of this chapter before presenting my conclusions. First, this essay does not seek to suggest that Native American speakers like Winona LaDuke invoke the Ecological Indian simply as a calculated means to allow themselves to speak on non-ecological issues. The Ecological Indian voice is not a charade or false front. Rather, I have tried to demonstrate a special condition: that those Native American individuals and communities already concerned with their biophysical environment—whether in and of itself or, as is more often the case, as an extension of a larger sociopolitical web—have, in recent years, enjoyed a special rhetorical avenue to success, unprecedented among modern-age indigenous populations.

Nadasdy's concerns that Native representatives working within the dominant Western system are still apt to be ignored or merely used to promote existing power structures bear our collective and concerted attention, but clearly some individuals, like LaDuke, are finding ways to harness the Ecological Indian identity and promote their own ends, and to publicly utilize and develop their own voices. In speaking, writing, filmmaking, and other modes of conduct, in the age of new ecology and value for TEK, the Native American activist can play the culturally sanctioned role well enough to parlay attention and authority into other issues for the Native community. From perspectives that understand ecological, social, and political concerns as intermingled extensions of one another, this development—when made articulately—can be quite fluid.

Second, this chapter does not seek to apologize for or sidestep the complex history of Native American and Euro-American peoples, nor the constraints and power dynamics that underlie the very modes of expression that Native American communicators put into use. It recognizes that the major part of Krech's argument holds great merit, that the Ecological Indian persona is indeed a construct influenced by Euro-American interpretations and motivations, and that even when a speaker like LaDuke asserts her personal and cultural investment in the environment that investment may have been shaped by a system that perpetuates the Ecological Indian stereotype. It is of no use for us to applaud the Indian who is ecological for speaking on ecological terms by simply (and erroneously) saying that he or she is manifesting an unfiltered, uncensored Native American voice without the effects of Euro-American distortion. In fact, the Indian who is ecological faces an intense double bind when speaking in the American public: she may resist dominant cultural expectations of environmental spiritualism and risk not being acknowledged,[11] or she may voice concerns for social, educational, medical, and economic policies through the Ecological Indian frame, all the while perpetuating a stereotype and perhaps contributing to a continued constraint on modes of Native American voice and public involvement. She may speak on environmental issues and gain attention for her tribe and cause, but may also wonder if, as Krech suggests, her internally felt commitment to and reverence for the environment is as much a product of the way dominant American narrative has influenced her identity as it is of her own design and experiences.

All voice is influenced by the discursive community in which it is used. That any of us, Native American or not, feel anxiety, duty, or concern for environmental sustainability, conservation, or responsible ecological practice, is a result of history and discourse that has encouraged us to think about and value such things in the first place. And, over the last hundred years, Native Americans—an already marginalized and subjugated group of diverse peoples—have especially been encouraged to embrace aspects of their personal experiences and cultural traditions that hold the natural world in special regard. It is at best naively presumptuous and at worst complicit in continued subjugation to suggest that when Native American speakers voice concern for the environment and TEK they are traveling the final emancipatory avenue toward autonomous self-definition and empowerment for Indian peoples.

Yet it seems equally presumptuous to tell any individual or collection of people that their personal values, their active voices, are merely the construct of an outside group. The Indian who is ecological can invoke the Ecological Indian both because dominant models expect and accept it *and* because it corresponds with deeply, internally felt personal values and anxieties. The construction of the Ecological Indian, in this regard, is more of what Richard A. Rogers (2006) would call a transcultural phenomenon, "created from and/or by multiple cultures, such that identification of a single originating culture is problematic" (p. 477). The formation of cultural symbols and practices and values, Rogers writes, "involves ongoing, circular appropriations of elements between multiple cultures, including elements that are themselves transcultural"

(p. 491), and this rule applies equally to the concept of the Ecological Indian. It is a product of Indian and Euro-American interaction. Cultural dominance does not negate the agency of seemingly subordinate groups (p. 478), even when dominant cultures influence the means by which marginalized peoples assert and reclaim their own voices (p. 495). Besides, of course, no culture or cultural group is static or stuck in place or time; all cultures are dynamic and self-sustaining, while always interacting with each other.

Exploring the ways in which existing discourses, imaginings, and representations influence internally felt personal values, drives, and identities will fuel the inquiries of communication scholars throughout the current century. We should ask, through attention to discourse and history and interactive social practice, what parts of Native American identity are guided by Euro-American ideals and how. This is the work of the future, and I suspect that we will find that no part of Native identity and voice is untouched by Euro-American interest and influence. Yet I maintain that this influence does not negate potential for Indian agency.

In the symbol of the Ecological Indian, we see a kind of common ground for discussion and understanding between Native American and Euro-American peoples. That ground is indeed marked by the dominant culture in America, but also by the Native American who freely enters into it. The dominant expectation for Ecological Indians overlaps with the voices of individual Indians who are ecological, and an arena for Native American voice in the dominant public is established. With access to this mutually sanctioned arena, the savvy Native American rhetor can seek acknowledgement, recognition, and respect on other fronts. This acknowledgement is by no means guaranteed, but the special authority given to the Indian when speaking ecologically opens a rhetorical window where few others can be found.

In this arena, Indians who are ecological are encouraged to speak and those who aren't particularly ecological are dismissed or discouraged, so there is certainly more critical work to be done. Future studies might consider the rhetorical strategies of "non-ecological Indians"—individuals who do not regularly refer to compassion for the natural world—for comparison. We can explore how the encouragement to celebrate ecological aspects of Native American cultures has influenced the shape and development of those cultures in recent years. Yet we can also recognize that the role of the Ecological Indian, as transcultural production, is in flux and is shaped as much now by those who invoke it as those who first perpetuated it. While it may have started as an imperialist Western construct, we can see that when ecology has been widely valued the concept has been appropriated by the very people it had one time disenfranchised. The Ecological Indian can indeed now be used as a tool for amplifying Native American voices in places where they have not always or often been heard before.

Special thanks to Matt Turner, Sean Kicummah Teuton, Jennifer Keohane, Kenneth Lythgoe, Kelly Jakes, and Ashley Hinck.

Notes

1. The speeches included in this study were selected because of their prominence in an online form. By being posted online, these speeches have reached a wider audience than LaDuke's more intimate speaking engagements. However, to ensure that my analysis of the online speeches applies to her other discourses as well, I have included in my consideration a speech that I personally attended (Laduke, 2011d), print materials (Anonymous, 1999; LaDuke, 2004), and websites representing LaDuke's own organizations (Honor the Earth; WELRP).
2. See Ono and Buescher (2001) and Rogers (2006; 2007) for a detailed discussion of the commodification of Native American symbols, peoples, and identities.
3. See Forsyth (1998), Acheson, Wilson, and Steneck (1998), Krech (1999) and Nadasdy, 1999.
4. Agrawal (1995) explains how distinctions between Western scientific knowledge and indigenous ways of knowing based upon supposed differences in subject matter, method, and emphasis on context are equally invalid.
5. See Paul Robbins' four theses of political ecology (Robbins, 2004, pp. 14–15).
6. See, for instance, Neumann (2004) for a description of how the U.S. Parks system actively displaced populations.
7. Krech gives several examples, including criticism of tribes that have sold land for strip mining. See Krech (1999, pp. 216–220).
8. See Silvern (2000) for an extended history.
9. See LaDuke (1999; 2004; 2008; 2011a; 2011b; 2011c; 2011d; 2012).
10. For a nuanced case example of this strategy compare LaDuke (2004) to Wilcox (2007).
11. "When Native Americans exhibit agency, acting in dynamic ways, actively appropriating non-Indian values and strategies, they risk being denied the status of real Indians by the dominant culture" (Rogers, 2006, p. 489).

References

Acheson, J. M., Wilson, J. A., & Steneck, R. S. (1998). Managing chaotic fisheries. In F. Berkes and C. Folke (Eds.), *Linking Social and Ecological Systems: Management Practices and Social Mechanisms for Building Resilience* (pp. 390–413). Cambridge: Cambridge University Press.

Agrawal, A. (1995). Dismantling the divide between indigenous and scientific knowledge. *Development and Change, 26(3)*, 413–439.

Anonymous. (1999). Native struggles for land and life: An interview with Winona LaDuke. *Multinational Monitor, 20*, 19–23.

Asen, R. (2002). Imagining in the public sphere. *Philosophy and Rhetoric, 35(4)*, 345–367.

Black, J. E. (2009). Native resistive rhetoric and the decolonization of American Indian removal discourse. *Quarterly Journal of Speech, 95(1)*, 66–88.

Brady, M. J. (2011). Mediating indigenous voice in the museum: Narratives of place, land, and environment in new exhibition practice. *Environmental Communication, 5(2)*, 202–220.

Carbaugh, D. (1999). "Just listen": "Listening" and landscape among the Blackfeet. *Western Journal of Communication, 63(3)*, 250–270.

Dickinson, G., Ott, B. L., & Aoki, E. (2006). Spaces of remembering and forgetting: The reverent eye/I at the Plains Indian Museum. *Communication and Critical/Cultural Studies, 3(1)*, 27–47.

Endres, D. (2012). Sacred land or national sacrifice zone: The role of values in the Yucca Mountain participation process. *Environmental Communication, 6(3)*, 328–345.

Forsyth, T. (1998). Mountain myths revisited: Integrating natural and social environmental science. *Mountain Research and Development, 18*, 107–116.

Grossman, R. (Director). (2005). *Homeland: Four Portraits of Native Action* [Motion picture]. Berkeley CA: Katahdin Productions.

Honor the Earth. (2011). *Honor the Earth.org*. Retrieved December 2011, from http://www.honorearth.org/

Krech, III, S. (1999). *The Ecological Indian: Myth and History*. New York: W.W. Norton & Company.

LaDuke, W. (1999). Land, life and culture: A native perspective. Speech, 27 October, Seattle, WA. Retrieved December 2011, from http://www.youtube.com/watch?v=UXA2zCfxxAw&feature=related

LaDuke, W. (2004). The political economy of wild rice: Indigenous heritage and university research. *Multinational Monitor, 25(4)*. Retrieved December, 2011 from http://www.multinationalmonitor.org/mm2004/042004/laduke.html

LaDuke, W. (2008). Indigenous thinking in a time of climate change. Speech, 31 March, Lawrence, KS. Retrieved December 2011, from http://www.youtube.com/watch?v=xG3xKVUAnU0

LaDuke, W. (2011a). The next energy economy: Grassroots strategies to mitigate global climate change and how we move ahead. Speech, 3 February, Seattle, WA. Retrieved December 2011, from http://www.youtube.com/watch?v=WEf-koY56-M&feature=related

LaDuke, W. (2011b). Keynote speech at the Mining Injustice Conference: Confronting corporate impunity 2011. Speech, 7 May, Toronto, Canada. Retrieved December 2011, from http://www.youtube.com/watch?v=Ydq_qFGt8j8&feature=relmfu

LaDuke, W. (2011c). Talk on Tars Sands extraction. Speech, 13 May, Portland, OR. Retrieved December 2011, from http://www.youtube.com/watch?v=BGx-ATJDZAM&feature=related

LaDuke, W. (2011d). Religion, faith and the land from a Native perspective. Speech, 6 October, Madison, Wisconsin.
LaDuke, W. (2012). Food sovereignty, biopiracy, and the future. Johns Hopkins Bloomberg, 12th Dodge Lectureship, delivered 10 April. Retrieved December 2011, from http://www.youtube.com/watch?v=Un1hx-n5Pcc
Lake, R. A. (1991). Between myth and history: Enacting time in Native American protest rhetoric. *Quarterly Journal of Speech, 77(2)*, 123–151.
Martin, C. (1978). *Keepers of the Game: Indian-animal Relationships and the Fur Trade.* Berkeley: University of California Press.
Miller, J. B. (1999). "Indians," "Braves," and "Redskins": A performative struggle for control of an image. *Quarterly Journal of Speech, 85(2)*, 188–202.
Morris, R., & Wander, P. (1990). Native American rhetoric: Dancing in the shadows of the ghost dance. *Quarterly Journal of Speech, 76(2)*, 164–191.
Nadasdy, P. (1999). The politics of TEK: Power and the "integration" of knowledge. *Arctic Anthropology, 36(1–2)*, 1–18.
Nadasdy, P. (2011). "We don't harvest animals: We kill them": Agricultural metaphors and the politics of wildlife management in the Yukon. In M. J. Goldman, P. Nadasdy & M. D. Turner (Eds.), *Knowing Nature: Conversations at the Intersection of Political Ecology and Science Studies* (pp. 135–151). Chicago: University of Chicago Press.
Neumann, R. P. (2004). Nature-state-territory: Toward a critical theorization of conservation enclosures. In R. Peet & M. Watts (Eds.), *Liberation Ecologies: Environment, Development, Social Movements* (pp. 195–217). London: Routledge.
Ono, K. A., & Buescher, D. T. (2001). Deciphering Pocahontas: Unpackaging the commodification of a Native American woman. *Critical Studies in Media Communication, 18 (1)*, 23–43.
Robbins, P. (2004). *Political Ecology: A Critical Introduction.* Malden, MA: Blackwell.
Rogers, R. A. (2006). From cultural exchange to transculturation: A review and reconceptualization of cultural appropriation. *Communication Theory, 16(4)*, 474–503.
Rogers, R. A. (2007). Deciphering Kokopelli: Masculinity in commodified appropriations of Native American imagery. *Communication and Critical/Cultural Studies, 4(3)*, 233–255.
Salvador, M., & Clarke, T. (2011). The Weyekin principle: Toward an embodied critical rhetoric. *Environmental Communication, 5(3)*, 243–260.
Silvern, S. E. (2000). Reclaiming the reservation: The geopolitics of Wisconsin Anishinaabe resource rights. *American Indian Culture and Research Journal, 24(3)*, 131–153.
Watts, E. K. (2001). "Voice" and "voicelessness" in rhetorical studies. *Quarterly Journal of Speech, 87(2)*, 179–196.
Wilcox, L. (2007). Going with the grain. *Smithsonian, 38*, 6.

4
Sustainable Advocacy: Voice for and before an Intergenerational Audience

Jessica M. Prody and Brandon Inabinet

Science has shown humans' abilities to affect the Earth in generations beyond their time. This awareness comes at a moment when massive social change pressures national and international speech within terministic screens (Burke, 1966, p. 45) that prioritize short-term economic growth and political needs. "Sustainability" and related terms restrain that way of thinking, prioritizing the long-term well-being of the environment, the economy, and society as they come into contact with one another. Even given this standard definition, though, the term is abstract enough to be manipulated. Some polluting corporations, for example, take advantage of it as a propagandistic buzzword. Even in more legitimate applications, the term is ambiguous. "Sustainability" sometimes denotes a field of arguments about justice within and across generations, while it also symbolizes a criterion for value judgments regarding present environmental claims and policy solutions. Groups and individuals attempt to find useful ideas of sustainability to solve entrenched problems facing societies. Yet they fail to account for the possibility that these global and intergenerational problems produce demands that cannot be addressed by existing models of communication. Just as Plato advocated philosophical dialogue as a model to combat the demagoguery of Athenian democracy, we contend that existing approaches to advocacy need to be re-envisioned for our present dilemma.

In this essay, we seek out a communicative model giving voice to these global concerns. The intractable differences regarding deployment of the concept "sustainability" show evidence of a global system that obscures injustices (particularly those faced by future generations). In such an environment, modes of argument and display should not be limited to mere opinion-sharing, but should use the whole range of embodied action for advocacy. Chaim Perelman and Lucie Olbrechts-Tyteca's (1969) idea of the Universal Audience usefully incorporates plural systems of human values and modes of appeal against any particular, rigid "rationality." Advocates would consider modes of argument and display deemed reasonable by an imagined audience—diversely composed and motivated toward adherence to claims they judge as legitimate in that context. In some situations, immediate advantage and the Particular Audiences' demands could still be at the forefront; but in questions of rights, fairness, understanding, and happiness, a reasonable attitude is to bias toward a community of others presently and into the future. In our review of Perelman's work, we find a gateway for looking toward the demands of sustainability. By modifying Perelman's theory to an Intergenerational Audience, we see a model of sustainable advocacy that helps create justice for future audiences. That awareness would help bring into being (1) an orientation of advocacy toward long-term planning (as product), (2) a sense of inclusive democracy (as process), and (3) consciousness of the current system (as critical praxis).

The concept of the Intergenerational Audience furthers the discussion of voice, because it does not replicate the frequent trap of locating voice either "within a speaker" (in sonority of voice, the narrative of their experience, the expression of their identity, all of which could be summed as the construction of "authenticity") or "outside the speaker" (as something imposed unconsciously by language, by looming and powerful constraints, or by the lack of listening or recognition). Eric King Watts (2001) notes that these definitions of voice often lead to thin theory (p. 105). In the first case, Watts' survey of the literature finds "voice" is nearly always found only as an uncritical or unexamined praise of the marginalized or silenced speaker whom the critic wants to support. In the second case, silent or oppressed voices are merely the side-effect of a system of dominance represented in language. Voice is merely a "citation," a systematically imposed space in which silence exists or voice is not heard.

Instead, we would agree with Watts that voice is located in relations, and go on to say that voice is located in the imagined articulation between audience, text, and speaker. Voice exists everywhere in the relationships between self and others, as the recognition of other and of self, but especially in those situations when an individual strategically takes up advocacy to realign these relationships. As critics engaged in sustainability, we seek to find voice when speakers engage ethical or environmental incongruities in relationships between audience, text, and speaker. In other words, we look for voice in the exploration of difference (i.e., generations, global positions, or dissonance in practices and stated ethics) as a way to advocate sustainability for immediate audiences while accounting for audiences that may yet exist.

In an analysis of a speech by Severn Cullis-Suzuki to the 1992 United Nations Earth Summit, we witness voice in the irony and shame of the audience listening to a child uttering well recognized truths hitherto squelched by the dominant global agenda. Cullis-Suzuki speaks to an imagined audience of the powerful as well as the powerless, of humans as situated not only in a conference hall but within the larger context of a natural environment, and needing basic capacities not just for that moment but for generations to come. That regard for plurality and justice, as constructed in situations and into the imagined future, guarantees voice in sustainable advocacy as a fruitful field of practice and criticism.

The difficulty of theorizing sustainable advocacy

Developing a theory of sustainable advocacy is made difficult by at least three constraints. First, as a discipline of theory and practice rhetoric has primarily defined itself by the here-and-now, even if case-based theory or features of advocacy can trend across particular moments. Rhetoric occurs in responses to specific situations and audiences, unlike transcendent philosophy or timeless literature (Wichelns, 1925). Second, the term "sustainability" is itself problematic because of its contested nature, as explored at the outset of this essay. Thus, introducing "sustainability" into yet another field of study carries the risk of generating vague connections without moving the concept forward. Finally, to remedy this, we root sustainability in intergenerational thinking, which asks the rhetorician to account for views imagined for future audiences. In this section, we work through these issues.

Communicative models for advocacy, such as Lloyd Bitzer's (1968) "rhetorical situation", stress "immediate" exigencies with messages adapted to the "particular" audience. Counter-models to this dominant mode have been notoriously incomplete. Take, for example, Foss and Griffin's (1995) proposal of "invitational rhetoric" as a subset of discourse at odds with the more traditional—in their words "patriarchal"—rhetorical models. They theorize dialogue that is cooperative rather than persuasively agonistic or antagonistic, "rooted in equality, immanent value, and self-determination" (p. 5). Although such values should be sustained through and among Intergenerational Audiences, we read this theory as individual-centered (emphasizing the relevance of personal opinion and experience), made effective through a rhetorical coding of self-denial (the appearance of a non-persuasive intent). In the particular series of issues we take as central—questions of sustainable justice—inconvenient truths often cannot simply be "posited," lest they be shut out entirely. Yet we are not opposed to the general idea that a normative model might exist that gains adherence through understatement and power incongruity. Indeed, Cullis-Suzuki's damning critique of the global order came from the polite questions of a small female child. Yet, a model of sustainable advocacy needs a wide latitude of means while including measures for persuasive validity beyond self and individuality, so that opinions are crafted and evaluated based on the significance to the debate and those affected by the outcome.

A model of sustainable advocacy must also contend with issues of the term "sustainability." Even scholars with similar value systems and environmental goals define the term differently, skewing any intervention with contested meanings. For example, some authors foreground the future effects of present action (see: Caney, 2009; Dudai, 2009; Page, 2007). They believe that contemporary contributions to climate change are wrong because the consequences for contemporary inaction will take significant tolls on future generations who must respond to an imperiled environment. Other authors foreground the link with the term "justice," explaining that one concept cannot be understood without the other (Holland, 2004). Thus, current issues of social justice are linked to sustainability, such as income inequality between the world's richest and poorest individuals and countries (Agyeman, 2008; Dower, 2004; Fitzpatrick, 2001). Still other scholars critique existing political structures and hypothesize institutional structures needed for just sustainability (J. Barry, 2006; Norton, 2004).

Across these divides, "sustainability" is taken to include the environment, the economy, and culture in its purview. Communication scholars identify this as problematic because of these divided connotations. Rhetorics of sustainability may diminish environmental concern in favor of the social and economic ones (Bricker, 2012), or emphasize environmental and economic issues at the expense of social justice (Kendall, 2008). In addition, the easy ability of unsustainable actors, such as commercial whalers (Cunningham, Huijbens, & Wearing, 2012), to appropriate the term deserves caution. We understand the trepidation, but we see the complexity of sustainability as its strength.

Sustainable advocacy must be focused on social and economic community in addition to environment and self, as both the material world (of natural resources, pollution, cultural artifacts, and money) and the immaterial world (of values, cultural practices, and social relationships) are caught up in the debates. Michael Jacobs (2004) argues that "[t]he disagreements over the 'meaning of sustainable development' are not semantic disputations but *Are* the substantive political arguments with which the term is concerned" (p. 26). "Sustainability" encourages dialogue and deliberation. It requires the weighing of values and consequences. And, most importantly for our purposes, we believe the term makes inroads toward enabling those acting and speaking under its guise to consider the communicative model we describe here, engaging future generations as audiences of thought and action. The messiness of sustainability invites intervention to participate in the "substantive political arguments" Jacobs traces.

Yet action regarding sustainability will be stuck in a quagmire unless communication turns toward intergenerational concerns. Conceiving of an imagined Intergenerational Audience is neither as hopeless as accepting all speech that seems appropriate to Particular Audiences as "appropriate" and "timely," nor as dreamy as imagining perfectly just or rational advocacy. This project is central to deliberative communities in the face of technological and private influences that narrow the scope and possibilities of ecologically sound rhetorical models (Goodnight, 1982; Schudson, 1995).

In privileging Intergenerational Audiences we follow scholars who focus on the material aspects of justice, such as natural resource allocation (Norton, 2004; Jacobs, 2004), as well as those who focus on the handing down of values and indigenous practices so that future communities can make decisions about their own welfare (Prugh, Constanza, & Daly, 2000). The various definitions of intergenerational justice have one commonality that we see as rhetorical and therefore intersubjective—they ask individuals to see themselves as listening to and speaking for and before community members in the present and those who will come after them. In this construction, rhetoric holds a tension within itself: to speak in a particular moment at the same time as premising arguments with an awareness of other moments, peoples, and impacts on future realities. In this sense our theory relates to those scholars (Watts, 2001; Couldry, 2009, 2010) who have examined the concept of voice, identifying a prophetic process that includes speaking and listening for the creation, maintenance, and evolution of communities.

If advocacy is to be sustainable, it must account for those who will exist after the immediate rhetorical situation becomes "effectively alleviated" or "positively modified," as Bitzer put it. Thus, by refocusing sustainable advocacy on intergenerational concerns, and by highlighting this intersubjective process of voice, we begin to find a term that can more easily be harnessed to strategic and critical practices.

Sustaining the Universal Audience

The concept we have found most useful in thinking through a sustainable model of advocacy is the Universal Audience, an idea also difficult for the field of rhetoric to integrate. As put forward in *The New Rhetoric: A Treatise of Argumentation* in 1958 (translated into English in 1969), the term does signal a transcendent ethical ideal that does not square with the situational and contingent realm of rhetoric. As James Crosswhite (2010) points out, many of the leading scholars who deal with the term find it at least partially incoherent (p. 431). Still, it marks one of the few attempts to manage situated rhetoric with a normative ideal. In the wake of the Holocaust and the midst of the atom bomb and space race, normative hopes for argumentation, pluralist rhetoric, and public action offered a powerful alternative to short-sighted rhetoric tantamount to authoritarianism.

Perelman and Olbrechts-Tyteca (1969) begin with the simple assumption that arguments are formed to address audiences for the purpose of gaining adherence (p. 14). The issue then is to find means to make those arguments better understood, both as to their status of truthfulness and their pragmatic uses. Perelman and Olbrechts-Tyteca thus separate Particular Audiences from Universal Audiences and the status of claims within each. Both are imagined by the speaker. The Particular Audience takes idiosyncratic and prejudicial commitments, less reliable for their truth-value but necessary for assent, as the groundwork for agreement or commitment to action. The latter imagined construct, the Universal Audience, seeks out claims that can attain assent that transcends any Particular Audience (p. 30). A speaker tries to "convince" the Universal Audience to generally shared principles and modes of reasoning, rather than "persuading" to temporary goals and prejudicial modes of argument (p. 35). Thus, the idea of the Universal Audience calls for limited agreement on what constitutes reasonability that sustains pluralism and continued debate.

What we see in Perelman and Olbrechts-Tyteca resonates with those scholars who have explored voice as a process. Couldry (2010) explains that voice is not just an act of speaking, but a process of recognizing our claims on each other as responsible human agents. For Couldry and Watts, voice is about not only offering one's perspective from lived experience but also listening and allowing oneself to draw connections to others. " 'Voice' is the enunciation and acknowledgement of the obligations and anxieties of living in community with others" (Watts, 2001, p. 180). In participating in this process of voice, community is acknowledged, constructed, reinforced, and changed as individual community members seek commonality amongst their differences. Such a transformation happens only through multiple interactions and moments of exchange.

In order to participate in this process, however, there must be some shared expectations of argument. Perleman and Olbrecht-Tyteca's Universal Audience suggests that while some goals or actions might be limited by national, political, or religious affiliations, the premises for argument and the modes of carrying out that argument ought to be, in important regards, shared (1969, pp. 110–111). Perelman and Olbrecht-Tyteca clearly put this regulative ideal in the mind of the speaker(s), not something one could access absolutely or perfectly.

This means that the Universal Audience is not utopian in its aspirations. The Universal Audience principle only suggests that the speaker gain adherence based on values shared beyond the Particular Audience, especially for what counts as evidence and as reasonable modes of reasoning. While a Particular Audience comes to a particular goal, the speaker would achieve "maximal intersubjective agreement" (Aikin, 2008, p. 11) within the parameters of his situation and imagined audiences. This does not guarantee a result of perfect equality or an ideally cosmopolitan mentality that will carry through generations; such a universal may not exist. In this sense, even norms for what constitutes good argument or truth will vary with passing speakers and generations.

One concern raised by this evolving norm of argument currently is the reliance on personal narrative within existing scholarship on voice. While we agree that identity formation is key to entering into a relationship with a text and audience, critics continually advise using it as *the* key to rhetorical success (Foss & Griffin, 1995). Personal narrative can be a powerful rhetorical mode for those outside the dominant political structure, but in most situations it sits low in a hierarchy of appeals. Indeed, voice could just as well come from data collection, expert testimony, or other means of evidence and argument. The weight of evidence is field-specific, and personal testimony cannot be professed as a transcendent good.

Emphasizing lived experience brings to mind global warming deniers saying that a really cold winter demonstrates global warming must be a hoax. Even in the aggregate, that sort of anecdotal testimony will fail to contradict worldwide monitoring by climate science experts, as well as the evidence that warming tends to affect some areas (such as the global poles, the Maldives, or the Nile river basin) more profoundly than others. It is especially dangerous to privilege personal experience narratives when individuals already feel better knowing less about the environment and latch onto private or religious experience as the infallible benchmark for certainty.

Notice that the criteria for Universal Audience acceptance also produce tensions that make outcomes merely tentative and not perfect. For example, one imagined version of the Universal Audience might be composed of the fittest people around the world to adequately judge the validity of a claim: scholars or scientists steeped in methodology. We would also, though, want to search out diversity in numbers of people for "maximal" agreement, not screening for access to education. The first imagined audience is an epistemological benchmark that could stifle discussion by what is presently conceived of as the best information available and the best means for processing it, while the second might endanger good knowledge by endlessly invoking individual and cultural differences. In both, though, arguers will find unity in an appeal to transcendent values beyond the limited capacities of the arguer's constraints.

A focus on intergenerational thinking also helps alleviate these pressures by directing the imagination toward the composition and values of future audiences. Knowledge and identity differences in the present would be subject to the ever-changing scripts and actions of public dialogue that create future audiences. Global warming science or any particular hierarchies about economy or society could disappear, except insofar as intersubjective agreement of most reasonable persons would allow—in their culture, from their perspective, and imagining for unlimited others. Over time, the Particular Audience becomes more aware of the temporal restrictions of their own experiences and environment, and with that knowledge, will adjust or transform those beliefs toward a quality that takes on others, both outside that context at that time and into the future. This gesture toward the transcendent and timeless is indeed not foreign to the rhetorical situation, as rhetoric meets the universal and particular in the moment. It fails when it does not engage a larger system of values than the matter at hand (Crosswhite, 2010, p. 438). The concept of the Intergenerational Audience pushes those values toward considerations of futurity.

There is, for this reason, a sense that sustainable advocacy follows no predictable or absolute standard. Rather, under the "convergence hypothesis" put forward by Bryan G. Norton (1991), discourse "can be pluralistic, even while seeking to develop a shared worldview—it may turn out that environmentalists have considerable freedom of intellectual choice regarding the philosophical aspects of a worldview" (p. 92). Sustainable advocacy is thus bigger than environmentalism, not in the sense that we supplement it with other primary modes of advocacy such as social justice or economic capabilities, but that all of these are made subordinate to the capacity to redefine the central issues and agendas at stake for future audiences.

Our model of the Intergenerational Audience also takes into account Perelman and Olbrecht-Tyteca's interest in local culture as necessary and indispensible. *The New Rhetoric* says, "We can only attempt to transcend the opposites or divisions we know about, thus: Each individual, each culture, has thus its own conception of the universal audience" (p. 33). The universal only finds expression through the voice of the particular. The literature echoes this sentiment on sustainable modes of judgment:

> Ecosystem health can be understood only in a cultural context—the land ethic is a locally determined sense of the good life, constructed with a careful and loving eye on the natural constraints imposed by the ecological and climatological context of that life.
> (Norton, 1991, p. 239)

While Norton is being direct as to what sustainable voices look like in the present context of his audience, he begs the audience to reconsider the "what" and "how" of sustainable advocacy—both the premises and modes of argument that will allow future justice through community participation. Recent research also asserts that perception of risk and reality, even when given scientific backing, has little adherence if not expressed in alignment with the norms and values of a particular community (Kahan, Jenkins-Smith, & Braman, 2011). Such speech in the current context has the capacity to excel by raising consciousness of incongruities.

Specifically, one clear way of demonstrating incongruity in the current context is to bring attention to the absence of intergenerational or long-term thinking in dominant economic or political planning. In a recently translated essay, Perelman and Olbrechts-Tyteca (2010) wrote about rhetorical attitudes toward time that support the integration of temporal issues into their conception of rhetoric. The authors suggest that most rhetoric takes on either a practical, diplomatic, or logical attitude (p. 332). A practical attitude, the dominant one, seeks short-term solutions, in a realist style, to matters that demand immediate answers. Secondly, and also common, the diplomatic attitude is to delay action, in bureaucratic style, on complex or irreconcilable disagreements. Finally, the "logical" style, as the authors privilege it, takes the largest scope of time possible relative to the matter at hand and takes into account the changes that might necessitate further public action. By juxtaposing this logical attitude against diplomatic or practical ones, advocates for sustainability today can raise consciousness regarding the incongruity of purpose and time.

To demonstrate this model, the second half of this essay applies these advancements to Severn Cullis-Suzuki's 1992 speech to the United Nations Earth Summit. This speech provides both a model for current sustainable advocacy and a moment to reflect on critical practices enabled by the Intergenerational Audience model.

Principles of sustainable advocacy

When she was nine years old, Severn Cullis-Suzuki founded the Environmental Children's Organization (ECO) in Vancouver, Canada, a program focused on children that educated families regarding local environmental issues and action. Members of the organization raised money to attend the 1992 UN Earth Summit (United Nations Conference on Environment and Development [UNCED]) in Rio de Janeiro, gaining the opportunity to speak to the body with members representing 172 world governments. The Summit was the birthplace of documents that were crucial to initiating new global cooperation on environmental issues, such as the non-binding "Agenda 21" document and the United Nations Framework Convention on Climate Change (UNFCCC), the non-binding precursor to the Kyoto Protocol. Although the proposals did not have the force of mandatory compliance, the meeting had a key role in framing international discussion for the coming decades, much like the Universal Declaration of Human Rights. Cullis-Suzuki would speak alongside the Dalai Lama and other important world figures at this landmark meeting.

Amidst the nine-day conference, Cullis-Suzuki's speech remained one of the most notable in scope and delivery, as well as most prescient to the convention given the synecdochal importance of children as representing future generations. At the Summit, delegates wept and gave a standing ovation to the address. By 1993, Cullus-Suzuki had been lauded for the speech by the UN Environmental Program's "Global 500 Roll of Honour" and published a book with Doubleday Press, entitled *Tell the World*, with practical family-oriented environmental advice. In more recent times, the speech has viral salience on YouTube as "The Girl Who Silenced the World for 5 Minutes," with over 25.8 million hits as of March 2013.

An epideictic text, her "Earth Summit Address" scolded world leaders for hypocrisy and pled for immediate adherence to action by raising awareness of the temporality of their planning. Cullis-Suzuki attempted to expose the diplomatic values of participants that, as Perelman and Olbrechts-Tyteca noted, soothed tension, delayed decisions, and asserted that resolution may only come with time itself. In place of diplomacy, she presented a litany of environmental issues with short-term easy solutions that could later be revised, adopting the practical attitude. But more importantly, she put forward, both through embodiment and emotional text, the importance of far-reaching, logical attitudes that try to adjust to intergenerational thinking. Because of this realignment of attitudes and incongruity, we find this text to be a paradigm for sustainable advocacy privileging an Intergenerational Audience.

Cullis-Suzuki began her speech by contrasting her attitude to that of the delegates, or adults more generally who traffic between the practical and diplomatic:

> Coming here today, I have no hidden agenda. I am fighting for my future. Losing my future is not like losing an election or a few points on the stock market. I am here to speak for all generations to come. I am here to speak on behalf of the starving children around the world whose cries go unheard. I am here to speak for the countless animals dying across this planet because they have nowhere left to go. We cannot afford to be not heard.

Prior scholarship on "voice" would draw our attention to her bravery in uttering these words, or the "voice" silenced previously. Our analytical perspective draws us to the scope and time frame she envisioned for her message, between the Particular Audience of imagined delegates and world leaders and Intergenerational Audience of "generations to come," affected by the natural environment. Cullis-Suzuki deftly crafted her perspective on the spatial and temporal relations between text, speaker, and audience (Particular Audience and Intergenerational Audience). The current issues, she suggested, were just symptoms of the larger catastrophes her generation, and by extension, the Intergenerational Audience, would face. Her youth becomes an embodiment of futurity. Against her body is the diplomatic body of already existing power relations. Cullis-Suzuki "interrupts the on-going flux of time and space by projecting the subject into it; the sound of [her] 'voice' bends space and time" (Watts, 2001, p. 184). She spoke herself into a particular moment of text and audience, as well as all the future generations to come.

As a speech of values, Cullis-Suzuki's "Earth Summit Address" does not offer the types of data or modes of reasoning that might be incumbent on adult orators to establish these seemingly apocalyptic ills. In the context of the larger convention, these ills regarding the interrelated issues of poverty and environmental crisis could be debated. This allowed Cullis-Suzuki a strategic advantage over other delegates, if she could forestall suspicions of childhood naïveté. However, by at least acknowledging the practical, and by doing so with the boldness of a female child willing to chastise the world community, Cullis-Suzuki embodied a deeper awareness of layers of time and meaning.

She set up this awareness by moving from shared childhood fantasies to concern for the future generations: "In my life, I have dreamt of seeing the great herds of wild animals, jungles and rainforests full of birds and butterflies, but now I wonder if they will even exist for my children to see." While clearly biodiversity is the focus here, the same argument from future generations could be imagined in any deliberative setting regarding future conditions and constraints. Cullis-Suzuki echoed the calls of many indigenous peoples to account for generations yet to come, and to take action with these generations in mind (LaDuke, 1999, pp. 198–199). Cullis-Suzuki's text is uniquely specific, though, in placing a sense of childlike wonder and curiosity as the premise for her argument. Mediating between the practical and logical, we can imagine the possibility of her or her children seeing a menagerie of animals and ecosystems. This advocacy thus built relationships between humans and between humans and the natural world—relationships that will continue to inspire advocacy for human and ecological plurality.

That relationship is imperiled, though. Cullis-Suzuki stated: "I used to go fishing in Vancouver with my dad until just a few years ago we found the fish full of cancers. And now we hear about animals and plants going extinct every day—vanishing forever." The loss of biodiversity is associated with the loss of familial tradition; as prophetic discourse often demonstrates, perils in the natural world form signs for perils in human relationships and meaning-systems. One mode of integrating nature is through non-Western practices of "listening" and giving voice to the inanimate and nonhuman world (Carbaugh, 1999). Schlosberg (2005) explains that such a process can also happen if we simply expand our notion of self-interest to understand that taking care of the nonhuman world means maintaining a livable world for humans and our future generations (p. 105). Cullis-Suzuki's speech helps us see that a world devoid of biodiversity will be a world severely crippled in cultural differences and experiences for humans, if not actual human life.

Cullis-Suzuki supported one of the larger, nascent themes of the conference relating local cultural practices with large-scale environmental shifts. Fishing is especially important as representative of a way of life for indigenous populations and food resources for the poor in her homeland. By the end of the speech, she connected these themes with the wealth disparity between the Global North and Global South, showing how the emphasis on consumption in the North prohibited it from behaving ethically with its Southern neighbors. By balancing the practical and logical and mediating between Particular Audience and Intergenerational Audience, Cullis-Suzuki enacted a type of rhetoric that meets the needs of current sustainable advocacy. We believe such rhetorical resources are at least implicit in most discourse. *Kairos*, or "opportune timing," is not sacrificed to the timeless or "universal," but instead is held in unique tension with long-term significance of the message. It would be unfair to expect advocates to construct messages that will always stand the test of time, or to even require them to imagine what exactly such ideals would be, but asking that they remind themselves their messages may contribute to the constraints of those in the foreseeable future is reasonable and prudent as a norm of rhetorical imagining and judgment.

The focus on poverty and difference between the Global North and South also highlights the importance of interconnectedness shared between concepts of a Universal Audience and Intergenerational Audience. Those in current positions of power may not be there tomorrow, and those marginalized now might soon move into powerful social positions. Cullis-Suzuki made this interconnectedness clear:

> Here, you may be delegates of your governments, business people, organizers, reporters or politicians—but really you are mothers and fathers, brothers and sister, aunts and uncles—and all of you are somebody's child. I'm only a child yet I know we are all part of a family, five billion strong, in fact, 30 million species strong and we

all share the same air, water and soil—borders and governments will never change that. I'm only a child yet I know we are all in this together and should act as one single world towards one single goal.

In identifying with her audience, Cullis-Suzuki began by acknowledging and respecting the elite positions of the Particular Audience, but she very quickly transitioned to interdependent roles foregrounded in feminist theory—positions that stress our relations as "mutually dependent beings" (Nussbaum, 1994, p. 190), possessing "relational ties that have implications for theorizing dependency and ethical responsibility" (Butler, 2003, p. 12). Cullis-Suzuki asked her audience to make decisions about policy based on more basic roles in this intergenerational framework than those that construct autonomy and power in the immediate present. While we cannot know for certain that family structures and ecological knowledge will remain constant throughout time (and in fact can assume they will not), we can assume that they will not change so drastically that the ideas of Cullis-Suzuki's message would be unfathomable, offensive, or exclusive for generations that follow. Familial relationships remind us of our own biological condition, tied not only to humans but also animals and plants. Moreover, she did this to suggest that these experienced roles ought to form a basis for shared judgment. All family members, including the lowly children, have agency and a role to play in this larger rhetorical process; no parent knows exactly who their child will become or under what environments they will live. Emphasis on intergenerational roles heightens the position of the politically marginalized in the Particular Audience.

Cullis-Suzuki does find herself at odds with our ethical model when she suggested that we "should act as one single world towards one single goal." Although supportive of the rhetorical act of unification, the speaker here lent credibility to audience suspicions of sustainability advocates who might want unreflective or coerced adherence rather than a process of argumentative contestation. Theorists of participatory and deliberative democracy instead put forward inclusive agonism (Gutman & Thompson, 1996; 2004; Young, 2000) that allows us to imagine an alternative rhetoric that provides the sense of equality Cullis-Suzuki invoked without the unity she imagined. As Robert Asen (2004) writes, "[s]triving for democracy constitutes a continuous process. Democracy signals an orientation toward action in various domains of human activity rather than a singularly delineated end" (p. 198). Unless Cullis-Suzuki implied debate as the "one single goal," something certainly not clear in the text, this rhetoric has dangers on both the practical level of losing political dissenters and the logical level of seeming to elevate unity over the argumentative plurality assumed in the Intergenerational Audience principle.

Cullis-Suzuki, though, balanced inclusivity, like temporality, against both Particular Audiences and Intergenerational Audiences. She began her conclusion: "Do not forget why you're attending these conferences, who you're doing this for—we are your own children. You are deciding what kind of world we will grow up in." Although she continued to address her audience as parents, she narrowed the audience in this section to those in attendance at the conference. While on the Intergenerational Audience level this return to political elites at the end might be exclusionary, it is important to be strategically oriented to the Particular Audience. Cullis-Suzuki, by imagining and discursively constituting both, asked that the Particular Audience members adhere to intergenerational evidence and modes of judgment.

This act of sharing frames of intersubjective reasoning and agreement with an audience gives way to the final aspect of sustainable advocacy as a current practice: consciousness-raising performance. As attested to by the immediate reaction and long-term durability of Cullis-Suzuki's address, deep attunement to logical and embodied incongruities between Intergenerational Audience and Particular Audience can shake the foundations of current consciousness. Karlyn Kohrs Campbell (1973) identifies consciousness-raising as a defining stylistic feature of women's liberation rhetoric. Although the consciousness-raising rhetoric she describes is historically and audience specific, family resemblances are clear. Like the women engaged in liberation rhetoric, individuals advocating in a sustainable fashion are entrenched in a culture that contradicts the very logic they are espousing. In that formation, the features of consciousness-raising include "affirmation of the affective, of the validity of personal experience, of the necessity of self-exposure and self criticism, of the value of dialogue, and the goal of autonomous individual decision making" (Campbell, 1973, p. 79). In an astute illustration of the Global North and Global South wealth gap, Cullis-Suzuki acknowledged her own participation in the unfair economic structure that contributes to pressing global environmental crises:

> In my country, we make so much waste, we buy and throw away, buy and throw away, and yet northern countries will not share with the needy. Even when we have more than enough, we are afraid to lose some of our wealth, afraid to share. In Canada, we live

the privileged life, with plenty of food, water and shelter—we have watches, bicycles, computers and television sets. Two days ago here in Brazil, we were shocked when we spent some time with some children living on the streets. And this is what one child told us: "I wish I was rich and if I were, I would give all the street children food, clothes, medicine, shelter and love and affection."

Cullis-Suzuki's repetition of "we" in the quotation above assigned herself blame for the privileged and selfish life of those in the Global North. She continued with this personal blame in her story of the Brazilian boy, as she constructed a contrast between this boy's imagined prosperity with the inability of empathy in her home country. She acknowledged what Nussbaum (1994) calls the "accident of where one is born," as she laments:

I can't stop thinking that these children are my age, that it makes a tremendous difference where you are born, that I could be one of those children living in the Favellas of Rio; I could be a child starving in Somalia; a victim of war in the Middle East or a beggar in India.

The questioning of privilege as a "right" among such vast inequalities and waste puts the speaker into a mode of consciousness-raising, examining and positing her own participation and entrenchment in an unsustainable system of entitlement liberalism. Speakers must, as Campbell (citing Sally Kempton) explains, "fight an enemy who has outposts in your head" (1973, p. 86). Cullis-Suzuki exhibited self-exposure, self-criticism, and the value of dialogue central to the consciousness-raising of sustainable advocacy. She did this by recognizing her place and participation in an unfair system, undertaking critique of her own action, and asking others to engage in the same process. Her text forced both speaker and audience into incongruous self-awareness.

Cullis-Suzuki encouraged dialogue through a series of rhetorical questions. The first of these concluded the story of the Brazilian boy. She asked, "If a child on the street who has nothing, is willing to share, why are we who have everything still so greedy?" After a list of lessons parents teach their children in kindergarten, such as "not to hurt creatures" and "share," she asked, "Then why do you go out and do the things you tell us not to do?" Her questions demanded a response to her entire message, lest her audience admit hypocrisy, asking UN delegates, "Are [children] even on your list of priorities?" The pointedness and prevalence of the questions throughout Cullis-Suzuki's speech suggests that she hoped to begin a transformative conversation, or make her voice unavoidable, in the minds of her audience. In this way, the questions model empathetic judgment that stemmed from her own consciousness-raising encounter with a boy from Brazil.

Sustainable advocacy does involve a different sense of judgment than was enacted in women's liberation movements. Questions of self-determination have to be modified toward future community determination, at the same time that individual "outposts" of unsustainable behaviors and values find resistance. Sustainable advocacy should guide communities to engage in deliberations about important questions, such as what material and human resources are needed to sustain a community and what values a community should preserve to ensure sustainable actions are undertaken in the future. Whereas the women's liberation movement sought to connect women based on their shared experiences of oppression, sustainable advocacy, at least in the developed West, must connect through intolerance of shared privilege and an individualism that comes at the expense of future community. The problems of hyper-individualism (McKibben, 2007) and media distortion (as expressed by Jurgen Habermas (1991), Nancy Fraser (1990), Gerard Hauser (1999), and Thomas Goodnight (1982)) provide significant barriers to new modes of judgment. Against these dominant modes, sustainable advocates will have to engage in projects that integrate entire bodies and lived experiences in applications across mediated and non-mediated landscapes. In this case, a 13-year-old girl scolding UN leaders performed intergenerational justice hard for the imagined audiences to ignore.

Into futurity and beyond

Contested theories of sustainability on every level show that theorists of rhetoric will have an important role to play in conversations of sustainability. In the present situation, on a practical, pedagogical and critical level, the time has come to get serious about ethical critique and discontinue the exaltations of orators for "finding voice" and shedding tears for "silenced voice." Instead, we need to construct a normative rhetorical model that lifts specific "silenced voices," particularly in debates over sustainability, because the effects of living unsustainably disproportionately threaten their livelihoods.

Considering the Intergenerational Audience for sustainable advocacy makes a significant contribution to existing models of voice, seeing inclusivity without overemphasis on the individual. Couldry (2004) argues that every space of political deliberation should enact "citizens' mutual respect for each other's inalienable capacity to contribute as agents of the public sphere" (p. 17). Our model of sustainable advocacy similarly encourages a pluralist, participatory mode of argument; yet, that capacity to contribute may not ensure entitlement to speak at any given moment or to weigh personal experience as the paramount act. Sustainable advocacy is a community-determined process of listening and speaking, with a view toward shared future constraints.

Moments may emerge where sustainable advocacy is not the appropriate model of voice for the moment, perhaps when immediate dangers outweigh concerns for future consequences (such as in moments where lives are threatened, immediate action is necessary, and time for communal deliberation does not exist); however, even in these urgent moments, the actions taken must be evaluated after resolution of the incidents in order to understand what future consequences or acts of intergenerational injustice may stem from them, so that negative consequences may be avoided or mitigated. Sustainable advocacy is a model primarily for communicating about issues that have implications for future generations. But the model also asks us to give attention to the generational contexts of the variety of issues public deliberations engage. Too often in contemporary public debate present-day concerns are privileged over creating a livable future. By highlighting the Intergenerational Audience sustainable advocacy challenges our privileging of the present and encourages a communication orientation that weighs present needs with those of future generations, requiring advocates, such as Cullis-Suzuki, to make decisions about when to target a Particular Audience and when to speak to the Intergenerational Audience. Our hope is that this model may foster greater intergenerational justice.

Communication scholars will need to critically engage their own participation in practices and theory that contribute to unsustainable societies. As scholars we need to consider what we value as successful and ethical advocacy. One of the difficult tasks of the field of communication is to produce theories that speak well both to the everyday practice and to the critical evaluation of discourse. Our hope is that the Intergenerational Audience might become such a term. However, there are some difficulties. The Intergenerational Audience principle shares in common with Universal Audience the weakness for critical interpretation of being an imagined construct of the speaker. Thus, like any other questions of intended audience, it is a concept that can be gestured toward but never fully known. Perelman and Olbrechts-Tyteca had some interest in situating the Universal Audience within rules of fairness and justice, but never fully explicated what constituted such a program. Alan Gross (1999) has used the Universal Audience principle to examine Lincoln's rhetoric as he imagined a community of "all rational people," and evaluates the speech against that criterion (p. 210). Similarly, we might just extend that claim for the Intergenerational Audience to say "an imagined community of all reasonable people, now and into the future." Yet the model remains open, allowing that the performative and inventional capacities of the speaker might be beyond the grasp of a critic. Even the emphasis on the mode of argument or the plurality of viewpoints included could bias the concept one way or the other. However, we do find from our reading of Cullis-Suzuki to an international, reasoning audience a few general insights for critique and practice of sustainable advocacy, such as:

1. Modes of reasoning and appeals should measure up between a "maximally intersubjective" level between the Particular Audience and Intergenerational Audience. To the critic, this means that sustainable advocacy ought to use argumentation under conditions that practitioners across borders would allow. Similar forms of reasoning and premises should be applicable in other contexts.
2. An advocate exercises a sense of timeliness that mediates between both situational needs of a Particular Audience and constraints of the Intergenerational Audience. To the critic of sustainable advocacy, this would mean a discourse that takes an attitude that does not discount the "logical" potential to the benefit of merely diplomatic or practical ends.
3. Advocates should be inclusive with speech that maximizes agency for others, given that the power arrangements of a Particular Audience might change in the future. To quote Carbaugh, "A proper affective attitude, in these moments, is humility within what is potentially a quite potent and sacred scene, asking pity for one's feeble self, seeking sympathy and compassion because one has been placed in the presence of possibly uncontrollable and overwhelming forces" (1999, p. 262). In contrast to Cullis-Suzuki, the Intergenerational Audience principle would ask that the audience not be merely the means to one end, but that discourse encourages continued argument and mediation.
4. Advocates should engage in "consciousness-raising performance" and judgment that uses embodied practice and critical thought to interrogate current systems and open up new modes of thought and action so that local and significant community life are made

possible. Given the current constraints on sustainable advocacy, the critic ought to be sure that messages work with rhetorical attitudes toward time to improve the context for Intergenerational Audiences, with speech that does not unnecessarily defer decision-making, nor reduce the complexity of interrelated human action to short-term solutions alone.

Prugh, Constanza, and Daly (2000) argue: "A more engaging politics will be necessary to achieve a sustainable world of our choice, as opposed to one imposed by nature's unpredictable responses to abuse" (p. 10). This type of political engagement can only occur if individuals imagine Intergenerational Audiences, audiences who are themselves capable of meaningful social action and sustainable behaviors. Thus, in our view, critics and advocates interested in sustainability will need to begin mediating in this way between transcendent and timely goals, inclusive and epistemic criteria for validity, and consciousness-raising and practical action. Such mediation can only happen in communicative action that locates voice as intersubjective. We hope our refinement and extension of Universal Audience theory, with the Intergenerational Audience model, will begin that paradigm shift, both within and beyond environmental discourses.

References

Aiken, S. (2008). Perelmanian universal audience and the epistemic aspirations of argument. *Philosophy and Rhetoric 43(4)*, 238–259.
Agyeman, J. (2008). Toward a "just" sustainability. *Continuum: Journal of Media & Cultural Studies, 22(6)*, 751–756.
Asen, R. (2004). A discourse theory of citizenship. *Quarterly Journal of Speech, 90(2)*, 189–211.
Barry, J. (2006). Resistance is fertile: From environmental to sustainability citizenship. In A. Dobson, & D. Bell (Eds.), *Environmental Citizenship* (pp. 21–48). Cambridge, MA: The MIT Press.
Bitzer, L. (1968). The rhetorical situation. *Philosophy and Rhetoric, 1*, 1–14.
Bricker, B. (2012). Salience over sustainability: Environmental rhetoric of President Barack Obama. *Argumentation and Advocacy, 48*, 159–173.
Burke, K. (1966). *Language as Symbolic Action: Essays on Life, Literature, and Method*. Berkeley: University of California Press.
Butler, J. (2003). Violence, mourning, politics. *Studies in Gender and Sexuality, 4(1)*, 9–37.
Campbell, K. K. (1973). The rhetoric of women's liberation: An oxymoron. *Quarterly Journal of Speech, 59*, 74–86.
Caney, S. (2009). Climate change and the future: Discounting for time, wealth, and risk. *Journal of Social Philosophy, 40(2)*, 163–186.

Carbaugh, D. (1999). "Just Listen": "Listening" and landscape among the Blackfeet. *Western Journal of Communication, 63(3)*, 250–270.
Couldry, N. (2004). In the place of a common culture, what? *The Review of Education, Pedagogy, and Cultural Studies, 26*, 3–21.
Couldry, N. (2009). Rethinking the politics of voice. *Continuum: Journal of Media & Cultural Studies, 23(4)*, 579–582.
Couldry, N. (2010). *Why Voice Matters: Culture and Politics after Neoliberalism.* Thousand Oaks, CA: Sage Publications.
Crosswhite, J. (2010) Universalities. *Philosophy and Rhetoric, 43(4)*, 430–448.
Cullis-Suzuki, S. (1992, June). Speech Before the 1992 UN Earth Summit. United Nations. Rio de Janeiro, Brazil. Accessed 19 April 2012 from http://www.youtube.com/watch?v=5g8cmWZOX8Q
Cunningham, P. A., Huijbens, H., & Wearing, S. L. (2012). From whaling to whale watching: Examining sustainability and cultural rhetoric. *Journal of Sustainable Tourism, 20(1)*, 143–161.
Dower, N. (2004). Global economy, justice, and sustainability. *Ethical Theory and Moral Practice, 7*, 399–415.
Dudai, R. (2009). Climate change and human rights practice: Observations on and around the report of the office of the high commissioner for human rights on the relationship between climate change and human rights. *Journal of Human Rights Practice, 1(2)*, 294–307.
Fitzpatrick, T. (2001). Making welfare for future generations. *Social Policy & Administration, 35(5)*, 506–520.
Foss, S., & Griffin, C. (1995). Beyond persuasion: A proposal for an invitational rhetoric. *Communication Monographs, 62(1)*, 2–18.
Fraser, N (1990). Rethinking the public sphere: A contribution to the critique of actually existing democracy. *Social Text, 25*, 56–80.
Goodnight, G.T. (1982). The personal, technical and public spheres of argument: A speculative inquiry into the art of public deliberation. *Journal of the American Forensic Association, 18*, 214–227.
Gross, A. (1999). A theory of the rhetorical audience: Reflections on Chaim Perelman. *Quarterly Journal of Speech, 85*, 203–211.
Gutman, A., & Thompson, D. (1996). *Democracy and Disagreement.* Cambridge, MA: The Belknap Press of Harvard University.
Gutman, A., & Thompson, D. (2004). *Why Deliberative Democracy?* Princeton, NJ: Princeton University Press.
Habermas, Jürgen. (1991). *Structural Transformation of the Public Sphere.* Tr. Thomas Burger. Cambridge, MA: MIT Press.
Hauser, Gerard. (1999). *Vernacular Voices: The Rhetorics of Publics and Public Spheres.* Columbia: University of South Carolina Press.
Holland, A. (2004) Sustainability: Should we start from here? In A. Dobson (Ed.), *Fairness and Futurity: Essays on Environmental Sustainability and Justice* (pp. 46–68). Oxford: Oxford University Press.
Jacobs, M. (2004). Sustainable development as a contested concept. In A. Dobson (Ed.), *Fairness and Futurity: Essays on Environmental Sustainability and Justice* (pp. 21–45). Oxford: Oxford University Press.
Kahan, D. M., Jenkins-Smith, H., & Braman, D. (2011). Cultural cognition of scientific consensus. *Journal of Risk Research, 14*, 147–174.

Kendall, B. E. (2008). Personae and natural capitalism: Negotiating politics and constituencies in a rhetoric of sustainability. *Environmental Communication, 2(1)*, 59–77.
LaDuke, W. (1999). *All Our Relations: Native Struggles for Land and Life*. Minneapolis, MN: Honor the Earth.
McKibben, B. (2007). *Deep Economy: The Wealth of Communities and the Durable Future*. New York: Henry Holt Company, LLC.
Norton, B. (2004). Ecology and opportunity: Intergenerational equity and sustainable options. In A. Dobson (Ed.), *Fairness and Futurity: Essays on Environmental Sustainability and Justice* (pp. 118–150). Oxford: Oxford University Press.
Norton, B. G. (1991). *Toward Unity among Environmentalists*. New York, NY: Oxford University Press.
Nussbaum, M. (Oct/Nov 1994). Patriotism and cosmopolitanism. *Boston Review*. Retrieved 17 August 2012 from http://bostonreview.net/BR19.5/nussbaum.php
Page, E. (2007). Fairness on the day after tomorrow: Justice, reciprocity and global climate change. *Political Studies, 55*, 225–242.
Perelman, C., & Olbrechts-Tyteca, L. (1969). *The New Rhetoric: A Treatise on Argumentation*. Notre Dame: University of Notre Dame Press.
Perelman, C., & Olbrechts-Tyteca, L. (2010). On temporality as a characteristic of argumentation. *Philosophy and Rhetoric, 43(4)*, 315–336.
Prugh, T., Constanza, R., & Daly, H. (2000). *The Local Politics of Global Sustainability*. Washington, DC: Island Press.
Schlosberg, D. (2005). Environmental and ecological justice: Theory and practice in the United States. In J. B. Eckersley (Ed.), *The State and the Global Ecological Crisis* (pp. 97–116). Cambridge, MA: The MIT Press.
Schudson, M. (1995). Creating public knowledge. *Media Studies Journal, 9(3)*, 27–32.
Watts, E. K. (2001). "Voice" and "voicelessness" in rhetorical studies. *Quarterly Journal of Speech, 87(2)*, 179–196.
Wichelns, H. (1925). The literary criticism of oratory. In A. M. Drummond (Ed.), *Studies in Rhetoric and Public Speaking in Honor of James Albert Winans* (pp. 181–216). New York: The Century Co.
Young, I.M. (2000). *Inclusion and Democracy*. Oxford: Oxford University Press.

5
Response Essay: The (Im)possibility of Voice in Environmental Advocacy
Danielle Endres

Few in the field of environmental communication would contest that society faces a number of impending anthropogenic threats to the planet. Currently national attention may be most focused on the approaching climate crisis due to greenhouse gas emissions and global warming. As of this writing, climate change and environmental activists from the Sierra Club to 350.org to Greenpeace to Idle No More are calling on their supporters to fight the proposed Keystone XL pipeline that would transport crude oil from Canada and North America to refineries in the Gulf Coast. Opponents suggest that the Keystone XL project not only facilitates extraction of "one of the dirtiest, costliest, and most destructive fuels in the world" (National Wildlife Federation, 2013), but also represents "game over for the climate" in terms of efforts to control CO_2 emissions to stem global warming (Hansen, 2012). Despite the scientific consensus on climate change and knowledge of what has to be done to prevent further global warming, climate activists have struggled to find a compelling voice that will persuade skeptical publics and intractable politicians to take action. More broadly, despite the findings that 64% of Americans believe that environmental protection is more important than economic growth (Leiserowitz, Maibach, Roser-Renouf, & Smith, 2011), environmentalism in its many manifestations has struggled and continues to struggle to find broad support for individual, institutional, and societal changes to mitigate environmental problems.

The chapters in this section suggest that there is a *crisis of voice* in environmental advocacy. Yet, voice has, for the most part, been overlooked in scholarship concerning environmental advocacy and activism. What became clear to me in reading these chapters, however, is the importance of the concept of voice for thinking about the future of environmental advocacy. Drawing from both Nick Couldry's (2010) and Eric King Watts' (2001) work on voice, the chapters highlight that voice is not just an individual process of accounting for one's experience but is an ethical process that takes place within communities. Efforts to better understand the voice(s) of environmental advocacy, such as these chapters, have significant scholarly and practical implications for environmental advocacy that seeks to create sustainable communities and respond to environmental crises.

The main question that arises for me from these chapters is seemingly simple and admittedly practical: what are the constraints and possibilities for the voice(s) of environmental advocacy? In the remainder of this chapter, I will reflect on how the chapters led me to this question and offer suggestions for further research that can enrich the examination of voice as a heuristic for environmental advocacy. I will begin with a discussion of the crisis of voice in environmental advocacy. Then, I will reflect on the signs of the crisis and potential solutions. Finally, I will introduce some topoi for future research that take up the possibility of voice for environmental advocacy.

The crisis of voice in environmental advocacy

In their identification of corporate ventriloquism, Bsumek et al. draw on Nick Couldry's (2010) argument that we face a crisis of voice due to the dominance of neoliberalism, a form of discourse that reduces politics to the implementation of market functioning. Couldry contends: "We are experiencing a contemporary crisis of voice, across political, economic and cultural domains, that has been growing for at least three decades" (p. 1). Reading the chapters in this section, particularly Bsumek et al.'s engagement with Couldry, and then reading Couldry's book refocused my thinking about environmental advocacy and environmental social movements to pay more attention to questions of voice. These chapters helped me to see a crisis of voice in environmentalism and environmental advocacy. By this, I mean that environmentalism—a voice made of many voices—is struggling to find a voice, or voices, of environmental advocacy that can effectively address the many environmental crises we face. Without discounting the successes of particular environmental movements and advocacy campaigns, the collective voice(s) of environmental advocacy have not yet achieved widespread success in persuading publics and governments to adopt the large-scale (local, national, and international) behavioral and political changes that are needed to ameliorate complex problems like climate change. Of course, this broad assertion is more complicated than simply finding the most effective messages, especially considering that there are many definitions of success and effectiveness within environmentalism. My assertion raises many additional questions about the role of voice in environmental advocacy. I intend to return to some of these complexities and questions later in the chapter. For now, I will start with explaining how the chapters in this section brought me to reflect on the crisis of voice in environmental advocacy.

From the chapters, we see both signs of the crisis and signs of possible solutions for addressing the crisis. Both Belanger and Bsumek et al. reveal signs of the crisis of voice for environmental advocacy by highlighting strategies that are used to perpetuate anti-environmental voices. Belanger reveals how Michael Crichton, who denounces climate change as a conspiracy created by politicized radical environmentalist scientists, has cultivated a voice that has achieved relative success in perpetuating climate skepticism. Bsumek et al. divulge corporate ventriloquism as a form of voice employed by corporations to uphold neoliberalism and argue for the continued necessity of coal to Americans. Corporate ventriloquism is not just an oppositional voice to environmental advocates pushing for decreased reliance on coal and other fossil fuels; more importantly, Bsumek et al. carefully demonstrate how it upholds the system of neoliberalism that inherently limits the possibility of voice beyond those that uphold the economic logics of neoliberalism. These two chapters, then, highlight powerful voices that stand in opposition to the voice(s) of environmental advocates and the challenges faced by environmental advocates seeking to persuade audiences of the need for adopting more environmentally sustainable practices. The relative success of these voices also gives hints about the failures of the voice(s) of environmental advocacy.

Beyond demonstrating what I argue are signs of the crisis of voice in environmental advocacy, the chapters in this section also offer potential solutions through examples of potentially consequential voices. Schmitt's examination of how the hegemonic construction of the Ecological Indian can be appropriated by American Indians as a way to find a credible voice in environmental decision-making reminds me of the value of re-appropriation as a tactic. As I will discuss in more detail later, this raises the question of how re-appropriation may serve as a response to the crisis of voice in environmental advocacy. On a broader scale, Prody and Inabinet's chapter calls for rethinking rhetorical theory to account for sustainability for future audiences. With the concept of an Intergenerational Audience, they offer a normative rhetorical concept that creates a voice for sustainability and future generations and an example that demonstrates the possibility of this concept as a tactic for environmental advocacy. In these chapters, we see potential solutions for addressing the crisis of voice in environmental advocacy.

This initial bifurcation of signs of the crisis and signs of solution provides a useful, albeit simplistic, starting point for examining the constraints and possibilities for the voice(s) of environmental advocacy. In the next two sections, I offer some further reflection that both builds on the chapters and draws from additional examples.

Signs of the crisis

A sign of the crisis of voice for environmental advocacy can be seen in the manufactured controversy (Ceccarelli, 2011) over climate change in the face of the unprecedented scientific consensus that global warming from anthropogenic greenhouse gas emissions is happening. Belanger examines how climate skeptics, such as Michael Crichton, have succeeded in creating a credible and successful voice "to mobilize doubt in the face of empirical evidence" (this volume, p. 47). He asks, "Why does Crichton (an expert science-fiction author) find a sympathetic audience in an arena beyond his area of competence when others with genuine expertise in environmental science do not?" (this volume, pp. 45–6). Through rhetorical maneuvering between patronizing and populist voices, Crichton is able to present climate change as a fabrication put forth by dogmatic scientists-turned-conspirators corrupted by radical environmentalist thinking. Crichton is not the only voice of climate skepticism. Political and corporate interests intentionally cultivated a controversy about climate change based in uncertainty (Banning, 2009), thus creating a voice that has since been taken up by numerous politicians, journalists striving to present both sides, and members of the general public to oppose a broad range of climate mitigation policies.

Even with a scientific consensus behind them, environmental advocates are struggling to find a voice that can effect change. As it is becoming increasingly clear that the voice of a scientific consensus is not sufficient to motivate action, what competing voices are available for resisting the seemingly compelling voices of uncertainty and skepticism? While Belanger's project is aimed at understanding how the voice of climate skepticism has been successful with certain audiences, this is only one step toward addressing the crisis of voice. If we are invested in scholarship that can have practical implications for advancing the voice of environmental advocacy, then we need also think about how to construct strategies that respond to manufactured controversy, climate skeptics, and other compelling anti-environmental narratives.

Bsumek et al. turn their attention to the way that corporations create a guise of voice that seemingly celebrates voice while actually undermining it. Corporate ventriloquism is just one example of the crisis of voice in the neoliberal age in which corporations hold powerful voices and have the power to create astroturf voices that bolster neoliberal values while simultaneously excluding the possibility of certain voices. Bsumek et al. explain "corporate ventriloquism is a way of recognizing voice under conditions of neoliberalism—but it is a voice that, in reinscribing neoliberal ideology, is not really a voice at all" (this volume, p. 38). Corporate ventriloquism, then, represents a potentially dangerous trend in the construction of corporate voices that then constrain the possibilities for environmental advocacy. Again, understanding the rhetorical workings and engaging in critical evaluation of corporate ventriloquism is valuable in understanding the constraints faced by environmental advocates and the strategies that have been successful in persuading certain audiences to continue to accept anti-environmental messages. Yet, there is also a need for research that takes our understanding of these strategies and uses them to better understand how the voice(s) of environmental advocacy might resist. While Bsumek et al. recommend foregrounding the value of voice as one way to resist corporate ventriloquism and other forms of appropriation, I wonder about the potential in appropriating corporate and neoliberal discourses and re-appropriating corporate ventriloquism and astroturfing into environmental advocacy. Certainly, recent emphasis on Natural Capitalism (e.g., Hawken, Lovins, & Lovins, 2000) and other market-based mechanisms to account for environmental externalities are examples of voices of environmental advocacy that work with the discourse of neoliberalism. Further research might help us flesh out the enabling and constraining consequences of environmental advocates working within the discursive systems that seem to be more successful with wider audiences.

Thinking about ways in which corporate ventriloquism creates a guise of voice led me to think about the guise of deliberation that occurs in many participation processes for environmental decision-making. Another sign of the crisis of voice in environmental advocacy can be seen in exclusionary processes that limit a diversity of voices from contributing to public deliberation about environmental controversies. Public participation scholars have detailed the ways in which decision-makers often create a guise of deliberation that gives the appearance of an open process of public participation while actually relegating public voice to pro forma public hearings once a decision has already been made (e.g., Depoe, Delicath, & Elsenbeer, 2004). J. Robert Cox (1999) goes further, arguing that public participation processes often construct certain voices as indecorous or inappropriate, which "functions to dismiss the epistemic standing of citizens and thus their ability to critique corporate or institutional claims" (p. 22). This indecorous voice, while allowed to speak, is constructed as always already out of place. Resultant decisions often claim to have given voice to environmental advocates opposing or seeking more sustainable solutions but ultimately go forward with plans that have potentially devastating environmental consequences for the local land community (Cox, 2013). Indeed, one of the fears of the opponents of the Keystone XL pipeline is that the State Department's public comment period will follow this model that seemingly values public voice while excluding it from the actual decision-making process. Through documenting the ways that decision-making processes exclude often marginalized public voices, these scholars highlight the institutional constraints on the voice of environmental advocacy.

All of these examples reveal how external forces—individuals, groups, corporations, and institutions that we might think of as outside the bounds of environmentalism—create conditions that limit the voice of environmental advocates and contribute the crisis of environmental advocacy. Though not always overtly antagonistic to environmental sustainability, these forces represent challenges that environmental advocates must contend with in constructing a credible, believable, and ultimately persuasive voice for enacting the behavioral and political changes needed to address the environmental problems from climate change to species loss to toxins. Identifying signs of the crisis and more specifically the specific strategies that are used to perpetuate anti-environmental voices and stifle environmental advocates is crucial. Yet, we need to move beyond critiques of the system and start thinking about how our critical attention to these strategies can have practical implications for developing tactics of resistance. For instance, researchers might ask: can or should environmental advocates appropriate the same or similar strategies that seem to be working to advance anti-environmental discourse?

As I reflected on these chapters and the crisis of voice for environmental advocacy, it became clear that the signs of the crisis couldn't be solely attributed to external forces. As environmental communication scholars continue to examine the role of voice in environmental advocacy, we must also consider the ways in which the voice(s) of environmental advocacy may be confused, constraining, or even possibly damaging to creating the conditions for substantial change. Indeed, Michael Shellenberger and Ted Norhaus' (2004) report on "The Death of Environmentalism" sparked an ongoing debate within environmental advocates about the appropriate voice of environmentalism and environmental advocacy. Calling out an older version of environmentalism for being too apocalyptic, defensive, and lacking compelling vision, Shellenberger and Nordhaus advocate for a change in voice, a more positive environmentalism based in an inspiring vision of hope and idealism that can reach broader audiences. Whether or not this new vision is the answer for environmental advocacy, it sparked an intense debate within environmental advocates about the appropriate or most effective voice of environmental advocacy. This debate reveals that the conversation cannot simply be about the external constraints on environmental advocacy. There are numerous challenges that come from within the broad ranks of the environmentalism movement. Environmental justice advocates have indicted the environmental movement as an exclusionary, white, upper-class voice (see for example: Sandler & Pezzullo, 2007). Further, when the voice of environmental advocacy is focused on individual consumer choices (such as buying a Prius or expensive organic products) does it not uphold the values of neoliberalism and consumerism that are arguably at the root of environmental crises? In other words, part of the crisis may stem from internal tensions within environmentalism about the causes, solutions, and definitions of environmental crises. While the chapters in this section move our thinking forward in terms of highlighting dominant voices that must be challenged, they do not delve into what might be harder to answer questions about negotiating the concept of voice within environmental advocates.

Potential solutions

What potential solutions for the crisis of voice of environmental advocacy can we glean from the chapters in this section? Schmitt's analysis of the Ecological Indian reminds us that re-appropriation has resistive and empowering potential, that people can find voice through positively adopting the names and frames used to marginalize them. As Schmitt suggests, "some Native Americans have found a means of re-appropriating their own image from Euro-American culture, making the imposed role of the Ecological Indian a source of empowerment and engagement in the public sphere" (this volume, p. 67). In what ways might environmental advocates find similar opportunities to re-appropriate stereotypes toward empowerment and resistance?

Thinking beyond the reclaiming of negative symbols into positive symbols, some environmental advocates are asking the question of how environmentalism might appropriate the successful framing strategies of conservatives in order to broaden its appeal. What are the opportunities for environmental advocacy to appropriate the discourses of neoliberalism and conservatism toward empowerment? Is it possible to respond to the external signs of the crisis mentioned above through forms of appropriation?

Though he does not use the term appropriation, George Lakoff (2010) argues that the environmental movement needs to reframe its message in ways that connect with core progressive values, effectively appropriating the value-based strategies that have been so successful for conservatives. This sort of framing shift is also at the heart of Shellenberger and Nordhaus' advocacy of a more positive idealistic environmentalism that connects with staple issues for U.S. Americans such as jobs and the economy. They argue: "Conservative foundations and think tanks have spent 40 years getting clear about what they want (their vision) and what they stand for (their values)... If environmentalists hope to become more than a special interest we must start framing our proposals around core American values" (2004, pp. 32–33). Can we view Lakoff, Shellenberger, Nordhaus, and others as a promising sign for the voice(s) of environmental advocacy? Have they hit upon a form of voice that has more potential to persuade broad audiences and effect change? Honestly, I am not sure. In line with Robert Brulle's (2010) criticism of these approaches, I am skeptical of this approach's ability to fundamentally challenge the logics of neoliberalism, consumerism, and progress that underlie environmental crises. And, yet, there may be pragmatic value in appropriating a type of voice—a value oriented voice—that seems to resonate with broad audiences. Again, this may reflect longstanding tensions within environmentalism about the best or most effective means to achieving change. While the chapters in this section do not offer an answer, they demonstrate the power of voice as a concept that allows for a deeper analysis of the tensions within environmentalism.

Another positive sign comes from reimagining rhetoric to account for the sustainability of future generations. Prody and Inabinet contend that "existing approaches to advocacy need to be re-envisioned for our present dilemma" of environmental crises (this volume, p. 88). They introduce the concept of the Intergenerational Audience, an extension of Perelman and Olbrechts-Tyteca's Universal Audience that accounts for an ethical commitment to sustainability, as a normative rhetorical model that gives voice to sustainability and future generations in contemporary deliberations about environmental issues. This model's potential is revealed in the case of the speech given by Severn Cullis-Suzuki at the Rio Earth Summit in 1992. Despite the power of the speech and its renewed popularity on YouTube and despite the promise in the Intergenerational Audience model, the challenge remains in the application of the model to contemporary environmental controversies. As Gardiner (2006) notes, climate change and other environmental problems are difficult to solve precisely because it is difficult for people to think in intergenerational terms. Future research on this and other models designed to give better voice to environmental sustainability will need to address the practical implications and the rhetorical consequences of messages created through this model. In short, will messages created using the Intergenerational Audience be successful in advancing the types of changes needed to achieve environmental sustainability?

Finally, the concept of voice itself holds promise toward imagining alternatives. Couldry's (2010) discussion of voice as value recognizes the importance of organizing society around the value of diversity of voices. Though it may be daunting to think about a way out of the neoliberal crisis of voice, future research should also engage with questions of how to create conditions that allow for the fruition of voice as a value. Bsumek et al. note, "Foregrounding the importance of voice, and reimagining social organizing principles around the importance of valuing voice, offers one way to connect advocacy with critiques of the ideological and rhetorical rationales that enable structural imbalances in the political economy of voice" (this volume, p. 40). The challenge for environmental communication scholars is to think both about the conditions that constrain voice and about practical ways to enable voices that can resist hegemonic forces and persuade to widespread collective action to address environmental crises.

What voice(s) for environmental advocacy?

While there are many themes and future directions that one might draw from the chapters in this section, they act as a starting point for a much-needed scholarly conversation about constraints and possibilities for the voice(s) of environmental advocacy. This conversation will require much collective effort from environmental communication scholars asking a variety of questions and examining a variety of case studies. Questions include:

- What are the tactics of voice that are most likely to influence the adoption of the broad changes that are needed to address current and future environmental crises?
- How do we define success or effectiveness for the voice(s) of environmental advocacy?
- What are the competing voices of environmental advocacy?
- Is environmental advocacy better suited to a multiplicity of voices or to being constructed univocally?
- To what extent should the voices of environmental advocacy promote structural and/or incremental changes? Individual and/or political solutions? Work within the system and/or advocate for radical change?

In this section, I propose a few topoi for thinking through current and future tactics for environmental advocates to address the crisis of voice.

First, as we engage in research to evaluate different forms of environmental voice, we must recognize that there will not be one single voice of environmentalism that acts as a panacea to the crisis of voice. Rather, as the chapters suggest, there are multiple forms of voice that can be used in particular situations. Steve Schwarze (2006) argues that in certain *kairotic* situations, environmental melodrama is an appropriate frame for mobilizing support for environmental causes. This might be taken as controversial advocacy considering that melodrama has been vilified as a destructive rhetorical frame. Indeed, as noted above, Lakoff, Shellenberger, and Nordhaus called for environmentalism to move away from negative "doom and gloom" messaging and instead promote positive messaging. However, a key point from Schwarze's argument about environmental melodrama is that there are different frames for different situations. The voice of environmental advocacy, therefore, should not be limited to one monolithic voice such as that advocated in *The Death of Environmentalism*. Rather, a rhetorical approach that takes into account situation, audience, appropriateness, and timeliness may offer the most flexibility for thinking about the voice(s) of environmental advocacy. Although Schwarze did not situate his argument within literature on voice, thinking about melodrama as one tactic of voice for environmental advocates is useful toward understanding the conditions in which successful voices are created and sustained. As such, a potential area for further research on voice and environmental advocacy is the continued analysis of specific tactics and forms of voice that might collectively guide environmental advocates in making future choices about how to construct credible and hopefully effective voices for change.

In addition to turning critical attention to cases and tactics, the concepts of appropriation and re-appropriation that Bsumek et al. and Schmitt raised in their chapters represent a second area for future thinking about the possibilities for environmental voice. To what extent can appropriation and re-appropriation act as consequential tactics of voice for environmental advocates? Beyond Schmitt's example of the power in re-appropriation of the Ecological Indian voice, we see another example in the re-appropriation of the indecorous voice. In an extension of Cox's (1999) argument that public participation processes hail certain populations as indecorous voices, Kathleen Hunt and Nicholas Paliewicz (2013) argue that publics should embrace the indecorous voice "as a *Kairotic* opportunity for rhetorical invention that renegotiates rhetorical possibilities of public participation" and "imagine the emancipatory potential of indecorum" (p. 3). Hunt and Paliewicz offer examples of publics using the indecorous voice to seize opportunities to disrupt, take time, and invent resistance to unjust participation models. The indecorous voice is a very different tactic than the positive voice advocated by Lakoff, Shellenberger, Nordhaus, and others. One embraces a seemingly negative label and the other attempts to construct positive framing. Considering differing rhetorical situations, however, we cannot say that one mode of voice is better than the other. Research that looks at the opportunities and consequences of appropriation and re-appropriation as tactics of voice for environmental advocacy can help advance our understanding of the conditions in which these forms of voice have the possibility of resonating with audiences and provoking pro-environmental action.

Returning to climate activism and the Keystone XL pipeline, a final area for research might be to think about the recent move to civil disobedience in the climate movement. This move to civil disobedience may be a variant of the indecorous voice in that it embraces a mode of voice that is considered inappropriate. Although civil disobedience is no stranger to radical environmentalism (e.g., Earth First, Sea Shepherds), it has not been traditionally associated with mainstream environmentalism. With the arrest of Tim DeChristopher—also known as Bidder 70—who engaged in civil disobedience by bidding on and wining 13 parcels of land in an oil and gas lease auction, a turn to civil disobedience was soon taken up by others in the climate movement. During the 350.org and Tar Sand Action Campaign two-week protest at the White House in August 2011, approximately 1200 people were arrested in a planned act of civil disobedience against Keystone XL. More recently, the Sierra Club lifted its 120-year ban on civil disobedience to engage in an anti-Keystone event in February 2013 that resulted in arrests (Sheppard, 2013). Bill McKibben (2013) of 350.org states:

> It's no accident that the merging fossil fuel resistance has sent so many people to jail in the last few years. That's because the overwhelming wealth of the fossil fuel industry means we can't outspend them; we need other currencies with which to work. Passion, spirit, creativity. And sometimes we have to spend our bodies.

As this new form of voice for environmental advocacy emerges, we have a significant opportunity to ask questions about the rhetorical consequences of the contemporary voice of civil disobedience in the climate change movement.

These are just a few foci that might guide future research on the crisis of voice in environmental advocacy. There are certainly many more. The importance of the chapters in this section is that they begin what I see as a fruitful and impactful conversation with theoretical and practical implications for the future of environmental advocacy. The concept of voice offers a valuable heuristic for thinking about environmental advocacy and environmental social movements because of its multifaceted focus on voice as both process and ethical commitment. Using voice, we can not only theorize ideal conditions wherein all voices have an opportunity to be heard but also better understand specific strategies and tactics of voice and their rhetorical consequences.

Conclusion

If environmental communication is a crisis discipline and a discipline of crisis (Cox, 2007; Schwarze, 2007), then one of the crises to which we should turn our attention is the crisis of voice in environmental advocacy. Scholars of environmental communication and in particular scholars of environmental advocacy, activism, and social movement should continue to pursue research that helps to promote and create space for environmentalism.

References

Banning, M. E. (2009). When poststructural theory and contemporary politics collide: The vexed case of global warming. *Communication and Critical/Cultural Studies, 6(3)*, 285–304. doi:10.1080/14791420903049736

Brulle, R. J. (2010). From environmental campaigns to advancing the public dialog: Environmental communication for civic engagement. *Environmental Communication: A Journal of Nature and Culture, 4(1)*, 82–98. doi:10.1080/17524030903522397

Ceccarelli, L. (2011). Manufactured scientific controversy: Science, rhetoric, and public debate. *Rhetoric & Public Affairs, 14(2)*, 195–228. doi: 10.1353/rap.2010.0222

Couldry, N. (2010). *Why Voice Matters: Culture and Politics after Neoliberalism*. Thousand Oaks, CA: Sage Publications.

Cox, J. R. (1999). Reclaiming the "indecorous" voice: Public participation by low-income communities in environmental decision-making. In C. B. Short & D. Hardy-Short (Eds.), *Proceedings of the Fifth Biennial Conference on Communication and the Environment* (pp. 21–31). Flagstaff, AZ: Northern Arizona University.

Cox, R. (2007). Nature's "crisis disciplines": Does environmental communication have an ethical duty? *Environmental Communication: A Journal of Nature and Culture, 1(1)*, 5–20. doi:10.1080/17524030701333948

Cox, R. (2013). *Environmental Communication and the Public Sphere*. 3d. ed. Los Angeles, CA: Sage.

Depoe, S. P., Delicath, J. W., & Elsenbeer, M. A. (Eds.). (2004). *Communication and Public Participation in Environmental Decision Making*. Albany NY: State University of New York Press.

Gardiner, S. M. (2006). A perfect moral storm: Climate change, intergenerational ethics and the problem of moral corruption. *Environmental Values, 15(3)*, 397–413. doi:10.2307/30302196

Hansen, J. (2012, May 9). Game over for the climate. *The New York Times*. Retrieved April 18, 2013 from http://www.nytimes.com/2012/05/10/opinion/game-over-for-the-climate.html

Hawken, P., Lovins, A., & Lovins, L. H. (2000). *Natural Capitalism: Creating the Next Industrial Revolution* (1st ed.). New York, NY: Back Bay Books.

Hunt, K., & Paliewicz, N. S. (2013). Are you listening? Indecorous voice as rhetorical strategy in environmental public participation. Presented at the 12th Biennial Conference on Communication and the Environment, Uppsala, Sweden.

Lakoff, G. (2010). Why it matters how we frame the environment. *Environmental Communication: A Journal of Nature and Culture, 4(1)*, 70–81. doi:10.1080/17524030903529749

Leiserowitz, A. A., Maibach, E., Roser-Renouf, C., & Smith, N. (2011). *Climate Change in the American Mind: Public Support for Climate & Energy Policies in May 2011*. Yale University and George Mason University. New Haven, CT: Yale University Project on Climate Change Communication. Retrieved 18 April 2013, from http://environment.yale.edu/climate-communication/files/PolicySupportMay2011.pdf

McKibben, B. (2013, April 21). A tale of two Earth Day heroes: Tim DeChristopher and Sandra Steingraber. *Grist*. Retrieved 26 April 2013, from http://grist.org/climate-energy/a-tale-of-two-earth-day-heroes/

National Wildlife Federation. (2013). *Tar Sands*. National Wildlife Federation. Retrieved 18 April 2013, from http://www.nwf.org/What-We-Do/Energy-and-Climate/Drilling-and-Mining/Tar-Sands.aspx

Sandler, R., & Pezzullo, P. C. (2007). *Environmental Justice and Environmentalism: The Social Justice Challenge to the Environmental Movement*. Boston MA: The MIT Press.

Schwarze, S. (2006). Environmental melodrama. *Quarterly Journal of Speech, 92*, 239–261. doi:10.1080/00335630600938609

Schwarze, S. (2007). Environmental communication as a discipline of crisis. *Environmental Communication: A Journal of Nature and Culture, 1(1)*, 87–98. doi:10.1080/17524030701334326.

Shellenberger, M., & Nordhaus, T. (2004). *The Death of Environmentalism: Global Warming Politics in a Post-Environmental World*. Retrieved 18 April 2013, from http://www.thebreakthrough.com/images/Death_of_Environmentalism.pdf

Sheppard, K. (2013, January 25). Sierra Club turns to civil disobedience to stop Keystone pipeline. *Mother Jones*. Retrieved 26 April 2013, from http://www.motherjones.com/blue-marble/2013/01/sierra-club-turns-civil-disobedience-stop-keystone-pipeline

Watts, E. K. (2001). "Voice" and "voicelessness" in rhetorical studies. *Quarterly Journal of Speech, 87(2)*, 179–196. doi:10.1080/00335630109384328

Section II
Voice and Consumption

6
Voices of Organic Consumption: Understanding Organic Consumption as Political Action

Leah Sprain

Introduction

The organic food industry has grown more than 20% per year since 1990 (Allen & Kovach, 2000) to account for 3.5% of all food products sold in the country (Organic purchases on the rise, 2009). As an article in the online environmental magazine *Grist* opens: "Organic has hit the big time" (Nijhuis, 2003). Growing in market share, organic consumption is economically significant, and, increasingly, part of a lifestyle movement in the United States. In recounting the traditional reasons why consumers purchase organic products (specific health concerns, better tasting, better for the environment) a *Better Nutrition* article adds one to the list: "organic are hip" (McIver, 2004, p. 58).

For some organics advocates, making organic food fashionable is the whole point. To other groups, such as the Organic Consumers Association and National Campaign for Sustainable Agriculture, organic consumers represent more than a trendy, lucrative market; they also represent political actors pushing for change. Organic food is considered the exemplary case of green consumerism (Allen & Kovach, 2000), using consumer power to push the system toward broad political goals such as sustainability or decreased dependence on fossil fuels.

More than supporting change via purchasing habits, the argument is that consumption *is* politics—the consumer as political actor. The idea that consumption is political behavior is not limited to colloquial arguments. Some political science and communication theorists also treat consumer behavior as political. Schudson (1998) argues that people "do politics" when they wear political buttons or teach their children about labels at the grocery store (p. 299). In *Political Virtue and Shopping*, Micheletti (2003) argues more specifically that consumption can, in some circumstances, be a political act in that it represents the potential to express a viewpoint, set the political agenda of other actors, and change the political landscape.

This argument hinges on a basic premise: consumption articulates a political stance. It expresses a viewpoint that can affect politics. It has a voice. But what is the voice? Where does it come from? Who controls it? And in this case, what does purchasing an organic product actually say?

Micheletti (2003) provides an answer to these questions. She considers organic consumption a "positive political consumerist endeavor" that uses certification labels to politicize products (p. 98). From this perspective, politics can be enacted in the economic realm through the purchase of a product that deviates from the norm, thereby politicizing the production process. Certification standards denote a consumer's preferences and function as the voice of consumption. In the case of organic consumption, this perspective suggests that purchasing organic products in the United States endorses organic production methods as defined by the United States Department of Agriculture (USDA). The USDA rules present an official definition of the term "organic"; only products that meet these rules can carry the organic certification seal. Therefore, it follows that purchasing a certified product represents demand for organic production methods as codified by the USDA.

Yet consumers report broader reasons for purchasing organic products. In surveys in Australia (Lockie, Lyons, & Lawrence, 2000) and Vermont (Wang & Sun, 2003), individuals expressed various motivations for purchasing organic products ranging from health, natural content, sensory appeal, and price to animal welfare, environmental protection, and political values (Lockie, Lyons, & Lawrence, 2002). Some of these values are written into organic certification codes; others, such as locovorism and health, are not. Research demonstrates that subtle interests of consumers are not encapsulated into certified purchases, which means preferences cannot simply be read off their market behavior (Newholm, 2000).

Rather than presume that the meaning of organic consumption is fixed by organic standards, scholars have explored the discursive construction of alternative meanings. Goodman (2000) provided a "snapshot" of the ways that "the discursive and material content of 'organic agriculture' is being contested and re-configured" (p. 215). Focusing on the organic industry, he demonstrated the tensions between industrial practices and the socio-ecological imaginary of organic farming.

Academics and practitioners have long been skeptical of dreamy visions of organic farming. Guthman (2004) provides a detailed deconstruction of the myth that organic food operates on a different logic than industrial agriculture. Tracing the evolution of organic lettuce from "counter-cuisine" to "yuppie chow," Guthman (2003) problematizes assumptions that organic food ought to be valorized as a political ideal. Likewise, Lockie, Lyons, and Lawrence (2000) focus on the construction of green foods, "analyzing the ways in which 'environmentally-related' symbols are used by a diversity of actors in the construction of green foods" (p. 316).

Several scholars have explored the political potential of organic consumption while simultaneously rejecting the notion that consumption patterns alone will change the agricultural system. Allen and Kovatch (2000) argue that organic markets provide the space and resources for social movement activity. As Howard and Allen (2006) summarize, "most analysts who study alternative food initiatives have concluded that eco-labels have inherent limitations and should not be relied on as the only strategy to achieve positive food-system changes, but nevertheless remain one effective tactic as part of a more comprehensive approach" (pp. 249–250). Building on this tradition, Boström and Klintman (2009) problematize a simplistic notion of political consumerism. Drawing on a broad literature review, they argue that conscious consumers are reflective and self-reflexive—political consumers do not blindly trust organic sources nor does political consumption blindly replace other forms of political behavior. Focusing on the ways in which consumers report making decisions, Boström and Klintman conclude that green labels do not adequately capture the "reflective trust" in the political potential of consumption—a trust that they consider to be hesitant and qualified. In fact, Boström and Klintman maintain that the real political potential of consumer behavior could be found in the sub-politics discussions about the limitations and benefits of labeling processes themselves. To further explore this hypothesis, they call for future empirical research that focuses on consumers' thoughts, reflections, and ambivalence about political consumption and voice.

This move to understand the multiple meanings of organic consumption is in line with the premise in food studies that food is thick with symbolism and multiple interpretations (LeBesco & Naccarato, 2008; Spurlock, 2009). Communication scholars studying food have focused on how language can change what food is and what we know about it (Mudry, 2010), how environmental messages are related to broader social justice and food security issues (Katz, 2010), and how a politics of the personal offers tactics for performing resistance through food consumption and production (Cooks, 2009). As Lindenfeld (2011) argues, communication scholarship on food has "the opportunity to unravel the socio-economic, political, and cultural complexities of lived experience and the ramifications of our everyday choices as citizens" (p. 19).

Rather than start with a theoretical stance that presumes a particular relationship between politics and consumption, I consider how individuals and groups construct and make sense of organic consumption by examining online discussions and materials promoting organic consumption. Through ethnographic discourse analysis of Usenet discussion groups and rhetorical analysis of promotional materials, I present three meanings of organic consumption: organic consumption as political action, organic consumption as tasteful consumption, and organic consumption as shutting up. This analysis reveals how a single social practice—purchasing organic products—functions as both voice and silence. Drawing on theories of voice, I argue that the political potential of these emic perspectives depends on who listens—and how—to the voices of organic consumption. Market forces tend to crowd out the political voices of organic consumption, favoring tasteful consumption over political demands. In the market, silence is not a form of resistance since it lacks political agency. The political potential of organic consumption then rests on the producers and organizations that attend to the voices of organic consumers seeking to express political values through consumption. This political potential is best realized through framing organic consumption as collective action by a social movement.

To establish my research orientation, I offer a conceptual definition of voice based on Hirschman's class work on exit and voice, and then compare this to contemporary conceptualizations for voice and silence. Then I briefly describe the methods grounding my ethnographic discourse analysis and rhetorical criticism. The next section introduces three voices of organic consumption: consumption as political action, consumption as tasteful consumption, and consumption as shutting up. In the discussion, I work across these voices of organic consumption to suggest theoretical implications for political consumption and voice.

Voice

Part of the theoretical justification for why a behavior should be considered political rests on the notion that it allows people to "express themselves politically" (Micheletti, 2003, p. 15). Transnational movements and individual actors "voice their views through their choices of products or producers" (Micheletti, Follesdal, & Stolle, 2004, p. x). This matches up with Hirschman's (1970) notion of political behavior as having a "voice" with which individuals "articulate" their demands. Hirschman (1970) describes voice as the attempt to change the policies and practices of the firm that one buys from or an organization of which one is a member:

> any attempt at all to change, rather than to escape from, an objectionable state of affairs, whether through individual or collective petition to the management directly in charge, through the appeal to a higher authority with the intention of forcing a change in management, or through various types of actions and protests, including those that are meant to mobilize public opinion. (p. 30)

This definition plays off its alternative—exit, the choice to stop buying a firm's products or leave an organization. In essence, Hirschman offers customers or organizational members two main options for influence: say something or leave.

The distinction between exit and voice is familiar. Typically, Hirschman's voice is enacted in interpersonal communication (i.e. offering a complaint about a business practice to a manager) or political behavior (i.e. public protest); exit occurs in the marketplace. Political scientists study voice; economists study exit. Economic organizations may not take voice seriously as those making complaints are seen as unrepresentative (Howard & Allen, 2010). Paradoxically, citizens have begun demanding voice in the marketplace, looking for ways to articulate preferences through consumer behavior (Micheletti, 2003). As such, political consumption can be classified as what Fairclough (2002) calls a "re-restructuring," a shift in the relations between different domains of social life under new capitalism, in this case re-restructuring the economic and political realms. From this perspective, politics can be enacted in the economic realm through voice, and consumers voice a preference for certain values by endorsing a production method that represents this change from the status quo.

Within rhetoric, voice is a contested term (Watts, 2001). Voice is invoked as a pre-discursive ability to vocalize (Appelbaum, 1990), a resistance to oppression (Huspek, 1997), the opposite of or antidote to silence (Bell, 1997), a socially located articulation (Dow, 1997), the language of a particular persona or speaking position (explicated in Jasinski,

2001), or even as the insertion of the scholar's perspective within academic writing (Strine, 1997). In a review of voice designed to clarify and specify the term, Eric Watts (2001) suggests another definition of voice as "a particular kind of speech phenomenon that pronounces the ethical problems and obligations incumbent in community building and arouses in persons and groups the frustrations, sufferings, and joys of such commitments" (p. 185). For Watts, voice is not unitary but a "happening" that functions from public acknowledgement of the ethics of speaking and the emotions of others, which is more than just the right to speak. The presence or absence of voice can be the object of inquiry; or voice can be intertextual, connected, and made evident by a critic (Watts, 2001).

For this paper, I use Hirschman's (1970) conceptions of exit and voice for initial analysis in the emic accounts of organic consumption. Then I return to rhetorical notions of voice to discuss the significance of multiple voices of organic consumption and the tensions between them. Rhetorical understandings of voice move beyond articulation to push the critic to account for the ethical and emotional event—the happening—created by voice.

Methods

To understand the potential voices of organic consumption, I look at two places: individuals discussing reasons for and meanings of organic consumption within Usenet groups, and promotional rhetoric encouraging organic consumption.[1] Analytically, I conducted an ethnographic discourse analysis of Usenet internet discussion groups and a rhetorical criticism of advocacy materials promoting organic consumption. To follow Burke (1966), these methods are teministic screens through which I can interpret meanings of organic consumption.

The ethnographic discourse analysis is grounded in the ethnography of communication (also called the ethnography of speaking), which is concerned with the "ways of speaking" in a particular culture or community by identifying the local means of speaking and their meanings to those who use and experience them (Philipsen & Coutu, 2005). An ethnographer of communication orients to local meanings by paying attention to the ways participants engage in communicative conduct, how they orient to their own and others' conduct, and how they talk about conduct (meta-communication). Speech Codes Theory argues that a researcher can look to speech itself for evidence of the speech codes of a community, which, in turn, suggest the significance of speaking and the available means of how interlocutors constitute the meanings of communicative acts (Philipsen, 1997; Philipsen, Coutu, & Covarrubias, 2004). In the most recent presentation of Speech Codes Theory (Philipsen et al., 2004), Proposition 5 maintains, "the terms, rules, and premises of a speech code are inextricably woven into speaking itself" (p. 62). This position supports examining communicative conduct as a method for determining a community's code. In turn, Proposition 4 states, "The significance of speaking is contingent upon the speech codes used by interlocutors to constitute the meanings of communicative acts" (p. 62). Thus, understanding a community's speech codes suggests the potential meanings of a particular act; "what a given behavior counts as, for a given receiver and interpreter of it, is contingent upon the speech code that the interpreter uses to constitute it as one sort of action or another" (p. 62). Given the possibility that consumption might be understood as functioning as speech (i.e. organic consumption "speaks" or "talks" or "communicates"), examining how participants invoke and orient to speech codes provides a way to understand (1) whether or not participants suggest or accept that consumption is communicative and (2) the particular meanings that a community might associate with this act.

I examined two Usenet discussion groups—alt.politics.greens and talk.environment—over a five-year period. Usenet was an early, open internet discussion forum organized by topic. Usenet archives remain publicly available. No demographic information is available about these groups since participation is anonymous, but hundreds of people regularly participated in these particular groups with thousands weighing in over the course of my five-year sample. I focused on these two particular groups due to the number of conversations (multiple posts between members) about organic food and the relevance of the overall board topic board to organic consumption. My data set included any post that included the term organic and that prompted conversations between multiple posters (as opposed to sharing a press release that gets no response). In total, I analyzed over 250 pages of text.

In each of the Usenet groups, I looked for empirical evidence of a speech code that might suggest a way of understanding organic consumption as strategic action by applying Speech Codes Theory (Philipsen et al., 2004) and the SPEAKING framework (Hymes, 1972). My objective was to discover whether participants accept consumption as a communicative act and, if so, the particular meanings associated with it. Thus my analysis argues how these particular communities discuss and understand organic consumption.

For the rhetorical analysis I started by looking to public arguments promoting organic consumption to look for possible voices of organic consumption, following Black's (1970) approach to look toward the text to guide interpretation. Rather than choose a single artifact as representative of promotional material for organic consumption, I read and gathered a wide variety of environmental and activism web sites, magazines, books, and promotional brochures, including: Organic Consumer's Association newsletters and website, the Sierra Club's True Cost of Food campaign materials, and the Center for the New American Dream website; articles from *Organic Style*, *Vegetarian Times*, *Utne Reader*, *Grist*, and the *New York Times*; promotional materials from commercial grocery stores and consumer and farmer co-ops. Clearly, this is not a representative sample of all writing mentioning organic food. In particular, arguments about buying organic food for health reasons are underrepresented. Nonetheless, this smattering of sources did allow me to get a sense of the dominant arguments supporting organic consumption, particularly those coming from advocacy organizations.

Drawing on a constitutive model of communication, I focused on how the narratives in these materials constitute or position an organic consumer. A constitutive model of rhetoric focuses on the effect of public discourse, "how subjects, personas, situations, and problems emerge as the effects of rhetorical practices" (Greene, 1998, p. 1). Instead of rhetors seeking to have the audience identify with a particular argument or ideology, a constitutive model of rhetoric argues that rhetoric creates a subject position for the ideal audience. In essence, the rhetoric constitutes an identity or subject position while simultaneously using that position as the basis for rhetorical address (Charland, 1987). Adopting this perspective on rhetoric positions the critic to focus on how a piece of rhetoric creates outcomes rather than attending to how a particular message responds to the rhetorical situation at hand. The power of a constitutive rhetoric is that it creates "a subject position to which inheres a set of motives that render a rhetorical discourse intelligible" (p. 147); rhetoric makes it possible for individuals to identify with a broader ideology in a particular way. Greene (1998) argues that "in a political sense, speech speaks for a particular subject, while aesthetically, it speaks into existence a figure of a subject" (p. 4). So while an effect of a constitutive rhetoric is to create a subject position, it simultaneously provides an ideology and set of motives that represent the audience that the rhetoric creates. This allows the rhetoric to become the public voice for an individual or at least representative of a public voice.

Voices of organic consumption

Through inductive ethnographic discourse analysis and rhetorical analysis, I identified three related ways that organic consumption is constructed and discussed: organic consumption as political action, organic consumption as tasteful consumption, and organic consumption as shutting up. Each of these types of consumption offers a narrative that provides an answer to the question of what purchasing an organic product actually says. This section presents each of these meanings of organic consumption before considering how they contribute to our understanding of political consumption and voice.

Organic consumption as political action

Environmental journalist Michael Pollan (2001) writes that "organic is nothing if not a set of values (this is better than that)" (p. 33). Yet a strain of organic discourse makes a stronger argument: purchasing organic products is a means of enacting political and social values.

Alt.politics.greens is a Usenet group that is concerned with "green party politics and activities." In this group, meanings of organic consumption are related to a broader code of communication. A set of rules and premises can be pieced together from how participants talk about talk: talk alone is not enough (to be considered a legitimate action). It is preferable to "do something." Actions "speak for themselves." When we "walk the talk," actions lead to "progress."

Part of this code for communication is a folk theory of strategic communication and political change. Government is not the ideal target; instead, "companies have to be persuaded." This happens through "consumers" "telling" them what they desire. Participants pair this preference with a certain means of persuasion: consumption. Participants express several reasons for this conclusion. For example, "The only way to hurt the evil people is via their profits.... they have no 'social conscience', it is merely a PR tool" (Curtis, 2001). Here the "evil people" or corporations are said to only value profits. Therefore, to influence them, a campaign must impact the bottom line rather than depend on companies having a "social conscience." In these cases, participants invoke consumption as having a voice; consumption provides a mechanism for "expressing" and "saying" and "telling" a preference. This speech code is consistently referenced in conversations about organic food and the rest of the alt.politics.greens group discussions. The voice of organic consumption as political action is even more fully developed in the advocacy materials promoting organic consumption.

Within the advocacy materials, consumers are repeatedly told to vote with their dollars. The Sierra Club's The True Cost of Food campaign (2005) begins with the premise: "We vote three times a day. With every meal we can choose to help the environment or to harm it." Consider the following excerpts from an online guide to eco-shopping (Landis, 2004):

> At this time of unrest and violence in the world... You can derive a sense of power from voting with your dollars for a cleaner, kinder, more peaceable world. By shopping these kinds of places and products, you are directing your valuable dollars—our most powerful vote—in a truly generous and constructive way. You are voting with your money—the only real vote Americans have, in my opinion— for a better planet and world... Every dollar you spend is a vote. You can express your concern for people, the environment, animals—all living things—and for peace and social justice, with what you buy. It DOES matter.

The repeated use of political language in terms of voting suggests that this behavior has political influence.

Here political language is employed metaphorically. "Voting with your money" not only invokes political language, but it relates economics and politics to empower the consumer. Within this position, the language of consumption stresses that consumers have choices. In this position, the power of a consumer stems from the ability to choose— and choose a competitor if a company does not practice certain values. This is, after all, the basis for Hirschman's (1970) notion of exit. But this passage argues that the power of the consumer is even greater when economic choices can result in a "cleaner, kinder, more peaceable world." Through the metaphor of voting, money becomes a tool that can cast influence, the mechanism that enables someone to voice their opinion and participate in decisions. Typically, businesses and corporations control the marketplace. In ideal notions of democracy, however, voting allows everyone to participate equally. Thus the metaphor of voting with dollars transforms the marketplace into a public forum where consumers have influence and power normally preserved for corporations. Just as elected officials are responsible to their constituents, companies are beholden to the consumers who purchase their products.

Political language emboldens consumers to see economic decisions as value-laden. Admittedly, economic behavior has long been recognized as influenced by individual values (Aldridge, 2003; Gutman, 1982;

Vinson, Scott, & Lamont, 1977). But this language is unique in that it conflates individual values and the ability to reach political ends. Thus the metaphor suggests that economic decisions are not just *influenced* by values but a way of *enacting* values to reach a particular end, similar to how Powell (2002) found vegetarians enacting movement values by not eating meat.

Stemming from the trope of political language that invokes "revolution" and "political change," political consumption arguments are paired with rhetorical tactics typically associated with identification and social movements (Stewart, Smith, & Denton, 2012). This includes naming an organic movement and recounting the gains of the organic movement through the use of the collective "we." A short film by the Organic Trade Association titled "Grocery Store Wars" (2005) sets up organic produce against conventional practices through a parody of *Star Wars*. A young "Cuke Skywalker" is instructed by "Obi Wan Cannoli" to lead the "organic rebellion." The market has been "taken over by the dark side of the farm. An empire of pollution and pesticides has ruthlessly conquered the market." Thankfully, there is "a new hope," "a growing resistance called the organic rebellion is fighting back." It's not just up to Cuke; instead "you" as the audience of the film can also change the market by what you choose at the store. This parody effectively plays off the good and evil dichotomy of *Star Wars* in crafting an environmental melodrama (Schwarze, 2006) that sets organic produce apart from conventional agriculture. In this allegory, the consumer holds the power for enacting change.

In addition to referring to organic consumers with the collective "we" or a label such as the "rebellion" or the "organic movement," organic discourse also mirrors the rhetorical strategies of unifying a social movement speech by dramatizing a problem and then calling for action through mythification (Smith, 2000; Vanderford, 1989). In the initial selection, this pattern is demonstrated in moving from feeling "saddened, frustrated, and helpless" about the state of the world to the ability to express concern for "peace and social justice, with what you buy." Here purchasing an organic product is offered as a particular act that enables individuals to transcend problems.

The Sierra Club (2005) not only employs political language to promote organic consumption, it argues that consumers are the only people who can change food systems.

Most of the world's problems can't be fixed by individual action... But ONLY consumers can affect the way food is grown and transported; this is an area where our actions make a difference. If we buy food that's grown sustainably supply will follow demand and it will become more available....

The way that "actions make a difference" is by adopting the position of a consumer. As consumers, individual action can affect change; conversely, other problems cannot be fixed by individual action.

By using political language and rhetorical tactics associated with movements, this rhetoric constitutes a political organic consumer. This consumer is motivated by social and political concerns, including contamination of groundwater, consumption of pesticides, and the political economy of agricultural production. Purchasing organic products is more than an endorsement of a particular system of production. Instead, organic consumption is a "vote" in favor of "concern for people, the environment, animals—all living things – and for peace and social justice." Simultaneously, purchasing an organic product is a way for an individual to be part of the "movement."

Organic consumption as tasteful consumption

Some promotional materials embrace consumption, glorifying organic consumption not for its political impact but its cultural currency. This functions as what Goffman (1951) called conspicuous consumption, purchases that send messages about the social status of an individual based on material items. This fits within a broader trend to use food as a marker of cultural capital (Opel, Johnston, & Wilk, 2010) promoted by a rhetoric of consumer choice (Freeman, 2010).

The now-defunct *Organic Style* magazine provides a prime example of appealing to and constituting a conspicuous organic consumer. The publisher Maria Rodale, who founded *Organic Style*, described her readership as "organic sensualists." "We started to see just a real trend to organic as a lifestyle. It was really more about pleasure, and about the senses rather than being puritanical—no sugar, no salt. It was about enjoying the sensuality and cleanliness that comes with it" (Williams, 2004). The magazine follows these guidelines by highlighting celebrity activity and self-interest. Here the focus is on pleasure—sensual appeal comes before political values (which may or may not make an appearance).

Each cover of *Organic Style* profiles a celebrity who embodies a desirable lifestyle. In December 2004/January 2005, Rosario Dawson takes a "healthy trip" to Australia. The article does quote Dawson discussing her eating habits:

I've always done a lot of alternative eating. I've taken raw-food prep classes. As a kid, I ate turkey bacon and tofu. I still eat fried chicken and rice and beans, but the burst of flavor in your mouth from organic fruit and vegetables is incredible—you forgot that anything can be that good. You can really taste the difference in an organic strawberry.

(Seo, 2005, p. 92)

Within this passage, the allure of organic food is based on its taste, marketed alongside other trendy food movements like raw foods. But the focus of this article is not even the food—it's the image. Glossy pictures of a woman who is "beautiful, natural, singularly spirited" imbue her endorsement of organic products with meaning. This meaning is based on lifestyle, an organic style. A woman with organic style is lively, adventurous, and takes time to pamper herself. But by starting with a celebrity (who also appears on the cover), *Organic Style* reinforces the notion that activism is not the central aspect of an organic style or even the motivating factor. Instead, an "organic style" is an aesthetic not a political stance. As a holiday gift guide of "gorgeous and green picks" promises, the gorgeous comes before and takes precedence over the green. *Organic Style* employs similar conventions to other women's magazines in that it provides information about where to purchase all of the featured products (as examined in Smith, 2000).

In a guide to purchasing coffee entitled "Coffee that's great to the last drop," the article employs highbrow consumer labels to describe particular products. Yet these labels are not restricted to organic and other certification schemes (what Micheletti (2003) calls "positive consumerist endeavors"). Instead, other consumer buzzwords are mixed into the descriptions. Coffee is described as "artisanal" yet "readily available," which suggests that this coffee has a desirable pedigree, but consumers do not have to work hard to purchase it. This balance appeals to consumer choice while setting apart certain products as particularly desirable. The focus is once again on the how these "flavorful beans" "give your taste buds a treat"; the by-product is that you "also" help the planet. Significantly, the article is addressed to the individual "you" who is put into the position of an individual experiencing sensual pleasures. This is different from notions of a collective "we" working together for social change; instead, the appeal focuses on how an individual gets to "choose."

Together this rhetoric constitutes a subject-as-consumer concerned with taste and comfort. This individual is encouraged to make decisions on what feels good. Unlike Smith's (2000) consumer motivated by lack who seeks new consumer goods, the conspicuous consumption appeal stresses the sensory appeal that can be gained through adopting an organic lifestyle. The appeal employs social status and celebrity to promote an organic style. It follows that consumers could purchase these products to reap the same benefits. Notably, purchasing an organic product remains a representation of status, not an articulation of a preference with an audience in mind. Tasteful consumption is promoted as a method of satisfying personal desire—not Hirschman's (1970) voice.

Consumption as shutting up

Both Usenet groups also include cases where organic consumption functions as exit. Individuals who do not share organic values (or are against an "organic movement") advocate consumption as a form of exit—a replacement to speaking, not a form of it:

> If you have something against big agribusiness, I have a simple solution for you: don't buy their product. Buy organic. It's as simple as that. But whatever you do, don't impose your moral standards on ME and say that I am not allowed to value the low-cost products of big agribusiness. Don't dare infringe on my right to choose what produce I want, even if the produce I choose has been treated with pesticides or toxins.
>
> (Mike, 2000)

In this excerpt, consumption is offered as the preferred means of expression and companies are the targets. But here buying organic is a "simple solution" that keeps "activists" (the participant's previous term) from seeking governmental decisions that might infringe on other personal choices. This implies that buying organic is a "solution"—but not a solution to political problems. Buying organic is a solution to personal dilemmas. This comment seems to invoke normative values of consumption (explicated in Aldridge, 2003) that consumption is about individual choice and individuals should be allowed to seek "low-cost products." "Moral standards" should not interfere.

Likewise, several conversations in talk.environment about genetically modified foods include a specific premise about how organic consumption might function as strategic action: if people are concerned about genetically modified foods, they should buy organic and be quiet. In these conversations, at least one individual would start pushing for labels about genetic modifications so consumers could "make informed choices." The reaction was that these complaints were not necessary. Individuals concerned about genetically modified foods should be satisfied with buying organic. Consider an example:

>Labeling advocate Jodette Green said that foods that have been

>genetically engineered need to be labeled.

????? They don't "need" to be labeled. Since the recent USDA standards for "organic" were approved, there is a label. Any food with the "Organic" label is free of GMO foods. All other foods can be assumed to contain GMO.

(Blair, 2001a)

This example suggests that organic labels provide sufficient information that no other action is necessary. But other participants take this idea further by following the suggestion to buy organic with "enough." Here the existing labeling system provides an alternative that makes further information in the form of labels unnecessary while also marking further conversation in the group unnecessary. "Foods labeled 'Organic' do not include gmo crops. If they do not have the 'Organic' label they can be assumed to include gmo. It is hardly necessary to have BOTH kinds of labels" (Blair, 2001b).

Although participants never used this language, this premise sounds similar to Hirschman's notion of exit. If individuals are not satisfied with a certain product (in this case, not satisfied with genetically modified food), they should stop buying it. Consumer choice is a road out of the political conversation about GMOs.

Discussion

This analysis provides empirical evidence of how organic consumption is caught within a re-restructuring (Fairclough, 2002) between the economic and political realms. Beyond Micheletti's (2003) argument that organic consumption is political because it politicizes the production process, some scholars argue that citizenship, politics, culture, identity, and economics are *always* intertwined (Lindenfeld, 2011; Miller, 2007). In the case of organic consumption, how are politics, economics, individual action, and collective action intertwined? I use the concept of voice to engage this question.

The application of Hirschman's (1970) scheme reveals fundamental tensions: the same act can be understood as both voice and exit, as political or consumerist. To what extent do these tensions decrease the political potential of organic consumption? Whereas Hirschman's (1970) conception of voice can help identify these tensions, it does not provide a means of resolving them. Instead, I turn to other theories of voice that help navigate the political potential of organic consumption within this shifting landscape.

Clair (1997) argues that voice and silence always exist in complex tension with each other. Care must be taken not to connect voice with empowerment without recognizing the power of silence as a form of resistance as well. Nonetheless, voice typically carries with it the potential for change as it involves speaking issues and concerns known to a broader public. Watts (2001) moves beyond voice as an articulation to argue voice entails ethical and emotional commitments that stem from speaking. This conception of voice moves beyond a focus on opportunities for speaking to consider what voice as a speech phenomenon creates and requires beyond articulation. Treating voice as a happening orients our attention to the communal features of discourse, particularly the ethical and emotional event created by voice.

Considering voice as a happening (Watts, 2001), I turn to Nick Couldry's (2010) discussion of voice in his attempt to theorize politics after neoliberalism. He begins with the premise that "having a voice is never enough" (p. 1); instead having an *effective voice* is a base concern. He argues that neoliberalism denies that voice matters since it privileges market forces above all, which has created a crisis of voice under which voice is frequently offered without any effect. Like Watts (2001), Couldry is concerned about the ethical commitments required by voice, but he focuses on developing a way of *valuing voice* that can create the foundations for social and political life by caring about humans' ability to give an account of themselves. His book places voice at the center of his social and political theory designed to get beyond neoliberalism.

Couldry's (2010) argument that neoliberalism denies voice by privileging markets poses important problems for the political possibilities of the voices of organic consumption. Couldry acknowledges that "what we do as participants in markets can, under some circumstances, contribute to voice, whether individual (the types of clothes I buy, the food choices I make), collective (fan communities or user groups) or distributed (consumer boycotts or buycotts)" (p. 12). The problem is that neoliberalism lacks understanding of the social processes underpinning voice that make it "an embodied process of effective speech" (p. 12). If markets are left to listen to the voices of organic consumers, the political potential of organic consumption seems small, particularly given the multiple voices of organic consumers that I've outlined.

Organic consumers attempting to articulate political demands to corporations can easily be co-opted or ignored if businesses elect to focus on conspicuous consumption concerns rather than political demands. Market logic, by default, ignores the political aspirations of organic advocates by rendering them consumer demand and nothing more. Moreover, buying organic to "shut up" about genetically modified food offers no resources for resistance. In this case, exit lacks political agency unless it is paired with someone articulating the motivations for exit, such as a boycott paired with a campaign that explains collective action. Despite Clair's (1997) warning to not simplistically equate voice with empowerment and silence with absence, in the marketplace silence does not offer resources for political resistance. Silent consumers forgo political agency. Instead, consumer behavior is likely to be seen by corporations and producers as loyalty to existing production methods rather than social movement demands.

Nonetheless, Couldry (2010) offers a social and political theory of voice that challenges organizations that ignore voice or require voice to come in commodified forms. To what extent could his theory offer ways to hear political voices of organic consumers? Couldry's (2010) vision of post-neoliberal politics is one that attends to all of the voices of organic consumption, which would render organic consumption an act of economic support and a political demand for change. Yet this vision requires organizations and individuals to attend to his richer conception of voice and its communal dimensions. In this respect, restructuring the political and economic realms seems like a site for political possibilities—but only if attention is given to protecting and listening to voices of consumption.

Empirically, this paper demonstrates that a Usenet group already has a code for understanding consumption as a means of strategic action that speaks and makes demands for a particular type of food system. Advocacy groups constitute organic consumers as political actors to similarly express themselves in the grocery store—acting politically as part of a collective movement. Yet these emic understandings of consumption as political action do not automatically translate into political change. Effective voice does not rely solely on those attempting to speak. These alternative voices remain easily co-opted by liberalism when we rely on markets—not people—to attend to voice.

The political potential of organic consumption hinges on two possibilities. First, producers, organizations, and corporations can attend to the voices of organic consumers. Across the nation, there are tens of thousands of food cooperatives, farmers markets, and organic producers already motivated to consumers' political voices. But the prospect of mainstream agricultural corporations listening for political demands seems remote since profit motivation makes it easier to cater to tasteful consumers. Motivating producers and organizations to listen for and attend to the political dimensions of organic consumers can best be achieved through collective action. Just as advocacy materials position consumers as part of a collective working for change, organic consumption stands its best chance of offering an effective political voice when it is viewed as social movement behavior. But for now, relying on markets to listen to political voice remains a way to curtail the political effectiveness of organic consumption. Whereas Couldry (2010) provides a theory of voice that makes recognition possible, this potential has not been fully realized.

Note

1. For a more detailed account of the methodology and descriptive results, see Sprain (2005).

References

Aldridge, A. (2003). *Consumption*. Malden, MA: Polity.
Allen, P., & Kovach, M. (2000). The capitalist composition of organic: The potential of markets in fulfilling the promise of organic agriculture. *Agriculture and Human Values, 17*, 221–232.
Appelbaum, D. (1990). *Voice*. Albany: State University of New York.
Bell, E. (1997). Listen up. You have to: Voice from "women and communication." *Western Journal of Communication, 61*, 89–100.
Black, E. (1970). The second persona. *Quarterly Journal of Speech, 56*, 109–119.
Blair, J. (2001a, January 31). Consumers question GE food [Message 2] sci.environment
Blair, J. (2001b, May 2). PBS: Harvest of Fear Tuesday, 24th April [Message 10]. talk.environment
Boström, M., & Klintman, M. (2009). The green political food consumer: A critical analysis of the research and policies. *Anthropology of Food*, S5. Retrieved from http://aof.revues.org/index6394.html
Burke K. (1966). *Language as Symbolic Action: Essays on Life, Literature, and Method*. Berkeley: University of California Press.
Charland, M. (1987). Constitutive rhetoric: The case of the people Québécois. *Quarterly Journal of Speech, 73*, 133–150.
Clair, R. P. (1997). Organizing silence: Silence as voice and voice as silence in the narrative exploration of the treaty of New Echota. *Western Journal of Communication, 61*, 315–337.
Cooks, L. (2009). You are what you (don't) eat? Food, identity, and resistance. *Text and Performance Quarterly, 29*, 94–110.
Couldry, N. (2010). *Why Voice Matters: Culture and Politics after Neoliberalism*. Thousand Oaks, CA: Sage Publications.

Curtis, R. (2001, March 27). Corporate front group attacks organic consumer association as "organic thugs" [message 2]. Alt.save.the.earth
Dow, B. (1997). Politicizing voice. *Western Journal of Communication, 61*, 243–251.
Fairclough, N. (2002). Language in new capitalism. *Discourse & Society, 13*, 163–166.
Freeman, C. P. (2010). Meat's place on the campaign menu: How U.S. environmental discourse negotiates vegetarianism. *Environmental Communication, 4*, 255–276.
Goffman, E. (1951). Symbols of class status. *British Journal of Sociology, 2*, 294–304.
Goodman, D. (2000). Organic and conventional agriculture: Materializing discourse and agro-ecological managerialism. *Agriculture and Human Values, 17*, 215–220.
Greene, R. W. (1998). The aesthetic turn and the rhetorical perspective on argumentation [Electronic version]. *Argumentation and Advocacy, 35*, 19–30. Page numbers in text reflect online pagination.
Grocery store wars. Join the rebellion (2005). Documentary film. Retrieved 25 April 2005, from http://www.storewars.org/.
Guthman, J. (2003). Fast food/organic food: Reflective tastes and the making of "yuppie chow." *Social & Cultural Geography, 4*, 46–58.
Guthman, J. (2004). *Agrarian Dreams: The Paradox of Organic Farming in California*. Berkeley: University of California Press.
Gutman, J. (1982). A means-ends chain model based on consumer categorization processes. *Journal of Marketing, 56*, 60–72.
Hirschman, A. O. (1970). *Exit, Voice, and Loyalty: Responses to Decline in Forms, Organizations, and States*. Cambridge, MA: Harvard University Press.
Howard, P. H., & Allen, P. (2006). Beyond organic: Consumer interest in new labeling schemes in the Central Coast of California. *International Journal of Consumer Studies, 30*, 439–451.
Huspek, M. (1997). Communication and the voice of the other. In M. Huspek & G. Radford (Eds.), *Transgressing Discourses: Communication and the Voice of the Other* (pp.1–16). Albany: State University of New York.
Hymes, D. (1972). Models of the interaction of language and social life. In J. J. Gumperz & D. Hymes (Eds.), *Directions in Sociolinguistics: The Ethnography of Communication* (pp. 35–71). New York: Holt, Rinehart and Winston.
Jasinski, J. (2001). *Sourcebook on Rhetoric: Key Concepts in Contemporary Rhetorical Studies*. Thousand Oaks, CA: Sage Publications.
Katz, R. (2010). You are what you environmentally, politically, socially, and economically eat: Delivering the sustainable farm and food message. *Environmental Communication, 4*, 371–377.
Landis, R. (2004). Guide to eco-shopping. Retrieved 11 December 2004, from http://www.bodyfueling.com/Eco-Shopping.html.
LeBesco, K. & Naccarato, P. (Eds.) (2008). *Edible Ideologies: Representing Food & Meaning*. Albany, NY: State University of New York Press.
Lindenfeld, L. A. (2011). Feasts for our eyes: Viewing films on food through new lenses. In J. M. Cramer, C. P. Greene & L. M. Walters (Eds.), *Food as Communication: Communication as Food* (pp. 3–22). New York: Peter Lang.
Lockie, S., Lyons, K., & Lawrence, G. (2000). Constructing "green foods": Corporate capital, risk, and organic farming in Australia and New Zealand. *Agriculture & Human Values, 17*, 315–322.

McIver, H. (2004, February). Organic hip. *Better Nutrition, 66,* 58.
Micheletti, M. (2003). *Political Virtue and Shopping: Individuals, Consumerism, and Collective Action.* New York: Palgrave MacMillan.
Micheletti, M., Follesdal, A., & Stolle, D. (Eds.) (2004). *Politics, Products, and Markets.* New Brunswick, NJ: Transaction Publishers.
Mike (2000, August 17). Give me a break about Stossel! [Message 8]. Alt.politics.greens
Miller, T. (2007). *Cultural Citizenship: Cosmopolitanism, Consumerism, and Television in a Neoliberal Age.* Philadelphia: Temple University Press.
Mudry, J. (2010). Counting on dinner: Discourse of science and the refiguration of food in USDA nutrition guides. *Environmental Communication, 4,* 338–354.
Newholm, T. (2000). Consumer exit, voice, and loyalty: Indicative, legitimation, and regulatory role in agricultural and food ethics. *Journal of Agricultural and Environmental Ethics, 12,* 153–164.
Nijhuis, M. (2003). Beyond the pale green. *Grist Magazine.* Retrieved 9 April 2005, from http://www.grist.org/news/maindish/2003/11/12/beyond/index.html
Opel, A., Johnston, J., & Wilk, R. (2010). Food, culture, and the environment: Communicating about what we eat. *Environmental Communication, 4,* 251–254.
Organic purchases on the rise (2009, June). *American Vegetable Grower, 57,* 26–27. Retrieved 29 August 2009, from http://0-web.a.ebscohost.com.libraries.colorado.edu/ehost/detail?vid=4&sid=40038bc6-9611-42ad-95de-78311375595d%40sessionmgr4001&hid=4206&bdata=JnNpdGU9ZWhvc3QtbGl2ZQ%3d%3d#db=aph&AN=43051610.
Philipsen, G. (1997). A theory of speech codes. In G. Philipsen & T. Albrecht (Eds.), *Developing Communication Theories* (pp. 119–156). Albany: State University of New York Press.
Philipsen, G., & Coutu, L. M. (2005). The ethnography of speaking. In K. L. Fitch & R. E. Sanders (Eds.), *Handbook of Language and Social Interaction* (pp. 355–379). Mahwah, NJ: Lawrence Erlbaum Associates.
Philipsen, G., Coutu, L., & Covarrubias, P. (2004). Speech codes theory: Restatement, revisions, and response to criticisms. In W. B. Gudykunst (Ed.), *Theorizing about Intercultural Communication* (pp. 55–68). Thousand Oaks, CA: Sage Publications.
Pollan, M. (2001, May 13). Naturally. *New York Times Magazine,* 30–42.
Powell. K. (2002). Lifestyle as rhetorical transaction: A case study of the vegetarian movement in the United States. *The New Jersey Journal of Communication, 10,* 169–190.
Schudson, M. (1998). *The Good Citizen.* New York: Simon & Schuster.
Schwarze, S. (2006). Environmental melodrama. *Quarterly Journal of Speech, 92,* 239–261.
Seo, D. (2005, January). Rosario gets down (under). *Organic Style, 11,* 90–99.
Sierra Club (2005). True cost of food campaign. Retrieved 22 May 2005, from http://www.truecostoffood.org/truecostoffood/takeaction.asp
Smith, C. D. (2000). Discipline—it's a "good thing": Rhetorical constitution and Martha Stewart Living Omnimedia. *Women's Studies in Communication, 23,* 337–366.
Sprain, L. (2005). Organic consumption as a means of political communication or social movement? An ethnographic and rhetorical exploration of the voice of organic consumption. Unpublished Master's Thesis, University of Washington.

Spurlock, C. M. (2009). Performing and sustaining (agri)culture and place: The cultivation of environmental subjectivity on the Piedmont Farm tour. *Text and Performance Quarterly, 29*, 5–21.

Stewart, C. J., Smith, C. A., & Denton, R. E. (2012). *Persuasion and Social Movements*, (6th ed). Long Grover, IL: Waveland Press.

Strine, M. S. (1997). Deconstructing identity in/and difference: Voices "under erasure." *Western Journal of Communication, 61*, 448–459.

Vanderford, M. L. (1989). Vilification and social movements: A case study of pro-life and pro-choice rhetoric. *Quarterly Journal of Speech, 75*, 166–182.

Vinson, D. E., Scott, J. E., & Lamont, L. M. (1977). The role of personal values in marketing and consumer behavior. *Journal of Marketing, 56*, 44–50.

Wang, Q., & Sun, J. (2003). Consumer preference and demand for organic food: Evidence from a Vermont survey. Presented at the American Agricultural Economics Association Annual Meeting, Montreal, Canada.

Watts, E. K. (2001). "Voice" and "voicelessness" in rhetorical studies. *Quarterly Journal of Speech, 87*, 179–196.

Williams, A. (2004, July 18). Going up to the country, but keeping all your toys [Electronic version]. *The New York Times*.

7
Vote with Your Fork: The Performance of Environmental Voice at the Farmers' Market

Benjamin Garner

Introduction

> Food is everything. You need food to live, and we take that for granted so much. For me, it's a conscious statement as to how I want things to be. I believe in dollar bills as little ballots, and you're placing votes. I think it's the root of everything. Especially meat.
>
> (Jennifer, Farmers' Market Patron)

Opel, Johnston, and Wilk (2010) write that "Food is the thin end of environmental awareness—a site where fundamental questions can begin to be asked, questions that often lead to challenging re-conceptions of our environments, our societies, and ourselves" (p. 251). If this is true, then farmers' markets represent a site of questioning, an alternate vision of food exchange, and a performative protest. The recent uproar last year over Chick-fil-A's stance on gay rights reminds us that food consumption or avoidance represents a form of political protest and civic activism. Salbu (2012) argued that that he and the rest of the gay community will "Vote with our feet" and boycott the fast food chain for their political stance. By contrast, Preston (2012) blogged about how overwhelming hordes of people showed up to answer Mike Huckabee's call for Chick-fil-A appreciation day. Similarly, farmers' market patronage often signifies political issues like sustainability (Alkon, 2008a; 2008b). My research examines how the performance of environmental values at the Downtown Lawrence Farmers' Market (DLFM) in Lawrence, KS exemplifies one manifestation of voice. Supporting the farmers' market represents a political voice that expresses concern about conventional agricultural practices. Placing one's body in a public setting provides a venue to voice environmental concerns to audiences like community leaders, other citizen-patrons, and the self.

Farmers' markets

The rise of the supermarket diminished the number of farmers' markets from the 1920s to the 1980s, but more recently farmers' markets have rebounded (Lyson, Gillespie Jr, & Hilchey, 1995; Trobe, 2001). Farmers' markets in the United States and Canada have grown from around 1,000 in the 1990s to over 7,800 today (Gowin, 2009; "Farmers market search," 2013). Estimates for farmers' market sales in 2005 were projected at $1 billion—a 13% increase since 2000 (Brown & Miller, 2008), even though that only adds up to less than 1% of the market share (Winne, 2008).

Farmers' markets provide several benefits: they shorten the food chain, are environmentally friendly, create a sense of community, and rebuild trust in the food supply (Alkon, 2008b; McGrath, Sherry, & Heisley, 1993; Retzinger, 2008; Robinson & Hartenfeld, 2007). Markets also represent an alternative to industrialized agriculture and the many negative practices associated with it (Kloppenburg, Lezberg, De Master, Stevenson, & Hendrickson, 2000). Prominent journalists like Michael Pollan (2007; 2008) have revealed how industrially produced food involves unknown chemicals, is shipped from afar, and harms the environment. Others cite unethical social dilemmas associated with large food producers, such as the mistreatment of workers (Stull & Broadway, 2004). In response to some of these issues, people want more sustainable and ethical food systems.

While data on patron motivation for market attendance has revealed that environmentalism ranks lower in getting people to markets than reasons like obtaining fresh products and meeting farmers, environmental issues serve as bedrock for the existence of alternative markets. Some analyses of farmers' markets reveal they are popular simply because they offer fresh produce and specialized products, and support the local economy (Andreatta & Wickliffe, 2002). Feagan and Morris (2009, p. 239) mentioned that 40% of respondents in their study cited a general concern for the environment, but this was "sporadically voiced." At the same time, others argue that environmental concern is a foundation for market existence. Winne (2008) contends that environmental and health concerns in the 1970s pushed customers to markets that offered more sustainable and healthier options. Retzinger (2008), too, argues that farmers' markets represent "a conscious choice to opt out of the industrialized model of food production" and that dollars spent at the market "constitute a form of a political action and power" (pp. 251–252). Despite not being ranked the highest motivating factor for market attendance, environmental concern remains a central value of markets.

Voice and political agency

To discuss voice means to look at how humans make sense of our lives through narratives. We examine who listens to that voice and notice what change occurs as a result. Couldry (2010) articulates a definition of voice that acknowledges voice as a process and a value. Voice as process emphasizes the ability to narrate about one's life, and voice as value reflects a culture's receptivity to voice as a form. Voice is tied to material and physical resources that affect its expression, and particular voices can be represented, denied, or ignored. Couldry (2010) emphasizes that the efficacy of voice depends on the audiences that listen, and consequently, audiences that control material resources and ignore voice have power to limit political efficacy. Watts (2012) argues that voice implies an audience that serves as a public validation. As a social experience, narratives require an audience to understand and recognize the voice of the oppressed, the hurt (Watts, 2012). However, does this mean that voice is only legitimated if all audiences respond a particular way? I would argue that we should be cautious not to restrict our definition of voice in such a way as to require a particular response from all audiences. Some definitions of voice prioritize the centrality of the hegemon for validating or denying voice, but those in power are but one of many audiences that might validate the voice. There are cases where the powerful exert a tremendous control over the oppressed and seem to render them voiceless. Others argue that even the lowest ranking person in a system of oppression can exert some influence over the powerful despite their seeming impotence (Giddens, 1979; 1984). In the fight against destructive agricultural practices, corporate agriculture will ignore advocates' complaints as long as possible to try to deny voice. Yet other audiences listen even if corporations attempt to deny voice— policy makers, citizens, and consumers. In successful cases, corporations change their policies.

Communication scholars have demonstrated that protests and consumption based performances can be effective at enacting social change even when corporations initially refuse to listen. Pezzullo (2011) articulates how in light of the powerlessness of government to regulate environmental issues, activist groups have sprung up to challenge the status quo through boycotts and "buycotts" (p. 125). Boycotts seem especially effective at speaking the language of corporate logic—by threatening the profits of a company through abstention. While corporations may be less motivated to change policies because of environmental or ethical concerns, they care where their consumers spend money. In addition to boycotts, a buycott encourages consumption at locations at particular times to publically reward companies that are more sustainable and ethical. Pezzullo (2011) illustrates that both boycotts and buycotts have been successful at using consumption to influence policy. Campaigns for environmental change also capitalize on publicity as a tool for success, in some cases to damage corporate reputation (Pezzullo, 2011) and in others to use celebrities to generate positive awareness (Todd, 2011).

Further, Zukin, Keeter, Molly, Jenkins, and Carpini (2006) argue that approximately one-third of American citizens reported using boycotts and another one-third said they used buycotts to influence public officials. This variety of consumer activism occurs within young and old demographics and reflects what Zukin et al. (2006) call indirect efforts that constitute "public voice" (p. 75). However, this is not to suggest that all activism is successful; rather, it is precisely the underdog, subordinate relationship some environmental groups perennially have with corporate and governmental power that makes effective campaigns impressive. The relationship between agency and structure demands that we examine each case contextually to see how agents who advocate for environmental causes act within and through constraint (Giddens, 1984). For this analysis, I define voice as a communicative act that necessitates an audience, is situated in particular contexts, involves a persuasive telos, and is characterized by performance.

Performance

I will explain why voice is inherently performative and define my use of the concept. For many I interviewed, farmers' market patronage becomes a reflexive, public, participatory support for alternative agriculture. Because many toss around the word "performance" without clearly defining it, it is often impossible to gauge the scope of its use. Carlson (2004) asserted that the concept is essentially "contested" (p. 73) and that scholars and lay people both use it differently. My use of the term "performance" rests on four key assumptions.

First, performance necessitates an audience. "Performance is always performance *for* someone, some audience that recognizes and validates it as performance even when, as is occasionally the case, that audience includes the self" (Carlson, 2004, p. 71). Intended or unintended audience depends on the context, but this aspect of the definition signals a communicative intent on the part of the performer to send some message—even if that is to the self. Thus, performance is intrinsically reflexive. Richard Bauman calls this quality "doubleness," a mental comparison of one's own actions with other potential actions—Schechner calls it "restored behavior," cognitive separation between one's mind and behaviors (Carlson, 2004). At the farmers' market, audiences can include the self, other citizens, city officials, and corporate agriculture.

Second, performances are often reflexive occasions when a community defines itself and reinforces belief structures through ritual enactments (Baxter & Braithwaite, 2002; MacAloon, 1984). During these times, communities use ritual to reinforce cultural norms and occasionally to challenge them (Bell, 1997). Collective cultural performances, spectacles, and festivals allow communities to keep the parts of their identity they like and omit the parts they dislike. Watts (2001) defines these civic voicings as moments when a group expresses collective frustration.

Cultural performances represent a communal reflexity—times when a community defines itself.

Third, physical, embodied participation is a critical part of performance. Faber (2004) writes of how Saint Orlan, a performance artist, uses her body in videotaped self-mutilating surgical performances to critique Western standards imposed upon women's bodies. Broadcasting these performances live in art galleries and over the internet, Orlan's body becomes a site of contestation—a fundamental part of her performance—and she communicates her message through the medium of her body. Others, like Bourdieu (1984), have deleneated how distinguished food consumption was a way for the upper class to create status and class markers. Hence, the physicality of the body allows for performance of values and ideas.

Fourth, performance has an enhanced public quality. The idea of public spectacle and the enactment in public arenas is a central tenet of performance theory (Bell, 1997; 2004). For example, Del Negro (2004) examined how modernity is performed in a small Italian village through a ritual promenading in the public sphere. Through this daily ritual, Italians perform their identity publicly—the purpose is to see and be seen. Spurlock (2009) argued that farm tours in North Carolina represent a "place-centered form of performative advocacy" in support of the family farm (p. 16). Performances have a public quality.

In sum, performance requires an audience, signals communal values and enhanced reflexivity, emphasizes the embodied nature of existence, and has a public quality to it. Based on this definition, I suggest farmers' market patrons partake in a performative event that sends environmental messages to audiences that include the self, other citizens and fellow market patrons, city officials, and industrial agriculture companies.

Setting

The DLFM currently has two main locations. The main Saturday market is located just off of the main downtown strip on New Hampshire Street between 8th and 9th Streets. This market seems similar to other markets that have been studied (Robinson & Hartenfeld, 2007) in supporting local products, sustainable agriculture, and a general sense of community values. The Saturday market is open from 7:00 to 11:00 a.m. Except during special events, vendors must follow a strict market code: all products sold must be locally grown within the region and produced by the vendor. Other stipulations exist, such as the requirement for vendors to obtain most if not all of the raw materials locally. Food cannot be shipped from afar, and while the market does not stipulate that food has to be certified organic, many farmers advertise things like "pesticide-free" or "hormone-free" to indicate the health and ecological soundness of their products. The market website reiterates the fresh and healthy theme, displaying photographs on the home page filled with fresh vegetables, families walking around, and farmers unloading products on a sunny market day. The market exudes a concern for local, fresh, high quality, and environmentally friendly food.

To frame the results that follow, I want to mention that there are options other than farmers' markets that are less public that allow people to use their purchasing power to influence agricultural practices but remain relatively anonymous to the public eye (community-supported agriculture, purchasing organic at the store, etc.). In *Food, Inc.*, the documentary on food and agriculture in the United States, Kenner (2008) contends that the consumer demand for organic products has caught the corporate eye and helps improve the system, even in large corporations like Wal-Mart. At the other end of the spectrum, community-supported agriculture (CSA) is an arrangement wherein a customer pays a set amount of money to a farmer, invests in that farmer, and in return receives a box of food products directly from the farmer based on the season.

At the farmers' market, the performance of values becomes a public promenade for all to witness because the market is located in the central, downtown location. Patrons' expression of voice extends to at least four possible audiences: (1) the self (notions of self-identity surrounding and ethics of sustainability), (2) other patrons and fellow citizens, (3) city officials that make decisions about local food market locations and policies, and (4) corporate agriculture. Conscientious consumption is important and can generate change (Pezzullo, 2011). I argue farmers' market patrons use consumption to effect change but also transcend consumption by performing in a public show of support and solidarity by attending the market. The market becomes a civically relevant public stage upon which environmental values are enacted.

Methods

I interviewed 19 people (ten female, nine male), representing 16 customers and three farmers from the DLFM in Lawrence, KS. Interviews lasted from 40 minutes to 75 minutes. I observed for 15 hours on 12 different market days. The University of Kansas Institutional Review Board (IRB) approved all methods and procedures. I gathered data from semi-structured interviews and participant observation. I began participant observation in the summer of 2011 and have continued through the spring of 2012. I attended market days, purchased goods, observed interactions, took photographs to assist memory, wrote fieldnotes, and visited farms. For this analysis, I focused on the 19 in-depth interviews. I formulated a research protocol as a framework and questioned participants based on these questions, but I remained committed to the principle of emergent data changing or adding to my list (Charmaz, 2006). Participants filled out a short, one-page demographic survey in addition to answering interview questions.

After interview data were transcribed, I used open and axial coding (Charmaz, 2006) that led to thematic analysis of the data (Braun & Clarke, 2006; Emerson, Fretz, & Shaw, 1995; Ryan & Bernard, 2003). I constructed themes based on criteria such as repetition, reoccurrence, local terms, and prevalence across contexts, based on its ability to illuminate research questions (Braun & Clarke, 2006; Emerson et al., 1995; Opler, 1945; Owen, 1984; Ryan & Bernard, 2003; Warren & Karner, 2010). Many of my participants made environmentally based comments during our interview, and I initially coded data into broad themes such as "the environment," "local," "politics," and many more before parsing out specific themes within categories. I took the initial theme concerning the environment and conducted digital pile-sorting to arrange ideas into more specific categories and discover nuances. Finally, writing up the results section was a crucial piece of the analysis process because it helped to break down categories even further and reveal which themes were well supported with data and which themes lacked solid evidence (Charmaz, 2006; Warren & Karner, 2010). For this analysis, I focus only on environmental motivations for going to the farmers' market, but it is important to note that these motivations are inextricably bound to many others. My claims and goals are not aimed at population representativeness. Rather, I hope to produce what Charmaz (2006) describes as theoretical representativeness, which fully explores and saturates conceptual categories within a given time and place. My analysis provides deep insight into patron values about alternative food and may help explain farmers' market growth.

Results

When I focused on the environmental narratives, two themes best encapsulated the data: (a) corporate responsibility for (b) agriculturally based environmental issues.

Anti-corporate, anti-grocery

Many expressed the view that corporate agriculture and traditional grocery stores are a problem. Practices in these institutions are perceived as irresponsible, and the farmers' market is considered a more conscionable place to buy food. This is reflected in a host of negative comments about traditional grocery stores. For example, Angela said:

> It sounds silly to say that we don't believe in the supermarket, but we don't believe in the supermarket. We don't shop at the supermarket really, except for pasta sauce and stuff like that. We go to the farmer's market to go grocery shopping.

When Angela and her partner Matt must go to the grocery store, they buy from a high-end local Co-op grocery store called the "Merc" (The Community Mercantile); it sells natural, organic, and health foods and is perceived as more ethical. Angela explained that in addition to shopping at the Merc, she and Matt are also a part of another buying cooperative:

> We go to the Merc. We're also a part of the New Boston Buying Co-operative, so we buy bulk stuff through them when it's cheaper than the Merc.

Angela and Matt try to buy as many of their groceries from the farmers' market as possible, which they said was upwards of half of their groceries during the summer. I asked them what their motivation for doing this was. Angela replied: "I think we both just believe it's better; it's better for us, it's better for local farmers, it's better for the environment, the food tastes better. I like being able to know." I asked further, "It seems like people have one reason or another: freshness, a dislike of big industry...." Angela answered: "I think it's all together." Matt added that they are "foodies" for moral and health reasons. For example, because of their concern for ethical meat production, they have strict standards for what meat they will eat. Angela said: "We will eat meat if we raise it and kill it. Or if somebody we know personally has raised it and killed it, but we don't do that very often." Matt and Angela do raise chickens in their backyard for the eggs, but they are primarily vegetarian on principle.

Angela, Matt, and I spent several minutes talking about how the farmers' market offers something different than a traditional grocery store because the farmers' market is more accountable and seems more environmentally and ethically friendly. Matt stated: "It's the whole reason farmers' markets exist. One reason. To give the sense of accountability to people." When I asked them which grocery store represents the opposite ethic of the farmers' market's accountability, Matt and Angela cited Wal-Mart. Matt called it "irresponsible" and Angela escalated the claim: "I honestly would go with a word worse than irresponsible—I don't know, I can't reach it." "Destructive," said Matt. Angela continued: "You've got irresponsible, and then you've got actually deliberately destructive." "Not deliberately destructive," corrected Matt. Despite their nuanced differences, this dialogue reveals how strongly Matt and Angela feel that many grocery stores shirk their environmental responsibilities. For them, the farmers' market offers a higher standard of food production and consumption.

Their narrative reflects the elements of performance previously defined: (1) they perform their support for the market publicly for all to see and for me as the researcher listening to their stories, (2) they use the market as way to define their identities and collective beliefs, (3) the embodied nature of their performance is reflected by the fact that they have radically changed their diet to consume more ethically, and (4) their actions serve as metaphorical markers of their beliefs through ritual performance of patronage and support. Patronage at the farmers' market is instrumentally important for them getting food to eat, and it also serves as an expression of their environmental values.

Similarly, Carlo attends the market because he thinks it is more environmentally friendly and goes against the ills of big agriculture. An undergraduate student who is a regular at the farmers' market, Carlo believes that farmers produce their food in ethical ways. He said: "When I buy a jar of honey, I imagine that it's done in a friendly way and in a way that would be akin to the narrative of the farmers that are trying to do things naturally." For Carlo, the market is an ideological narrative that people participate in. He stated:

> You feel like they're [farmers] tapping into what is tradition instead of this new industrial, with the industrial strains of hormones or things like that.

The farmers' market narrative runs counter to the convenience-based way Carlo used to shop for food. He said: "It definitely contrasted with what I've been, what I used to eat. I used to just get what was convenient." Carlo realizes he partly buys food at the market to assuage guilt about unethical food, and farmers are "all in tune with this thing that is supposed to make you feel good about what you eat—like you're doing the right thing." Concern for ethics and the environment, according to Carlo, is part of the farmers' market narrative. Now, Carlo critiques industrial farming and hormonal alteration and feels that it is important to support local farmers who practice ethical production. Similarly, Irene's comments connected with this anti-industry sentiment. She said:

> There are people who support it [farmers' market] specifically for the idea that it is not a large corporation. It's involved; that it is just a group of independent people.

Similarly, Peter revealed his disdain for large agricultural corporations when he discussed the assumptions he makes about the farmers' market: "You're buying directly from a real person that grows it [food], and I assume they are not agri-business in disguise." Lisa, a scientist and inspector in the solid waste industry, sees many environmental issues at her work. She said: "I'm not a big fan of corporate agriculture. I don't think it's really good for the planet."

Jaclyn, too, revealed unease with commercial farming. She said:

> I think there's a sense of supporting people that are actually farming themselves as opposed to big commercial operations. Seeing the people who benefit from it...they tell you when they bought a

new tractor [or] they're getting a new truck. They've got their kids there—as opposed to this is going to some corporate thing.... I don't really know what that corporation is spending their money on, what their farming practices are. Do I agree with the way they are conserving their land and water and... chemicals they're using? I won't ever know that when I buy that at the store.

Jaclyn's comments reflect the pros of supporting a local farmer and the environmental cons of purchasing food at a typical grocery store.

Of course, not everyone who attends the farmers' market fears conventional grocery stores. While some people preferred the farmers' market or the Merc, others mentioned that they still shopped at grocery chains. Chris, an information technology support specialist, has been going to the farmers' market about once a month for the last three to five years with his wife and kids. He and his family shop at several chain grocery stores including Dillons (Kroger affiliate) and Target. He said: "Yeah, I have complete faith in what I buy at the grocery, just like I do at the farmers' market." Not everyone distrusts grocery store products and practices.

In sum, these narratives critique corporate agriculture and support principles espoused at the farmers' market. Voice reflects the emotional dimension that Watts (2001) discusses when he says that voice is "a happening that is invigorated by a public awareness of the ethical and emotional concerns of discourse" (Watts, 2001, p. 185). Voices of many patrons at the market express frustration, fear, and environmental concern. Voice in this context comes to represent the public display of values—the "sound of specific experiential encounters in civic life" (Watts, 2001, p. 185). Participation in the farmers' market, for some, is civically minded food activism.

Environmental concerns

The second major theme reveals an explicit concern for the environment and explores the harmful practices of corporate agriculture. Customers want more environmentally friendly food, and the farmers' market meets that need. Within the "environmental concern" theme, five overlapping sub-themes emerge as areas of concern in patron narratives: (1) farming methods, (2) pollution, (3) unethical practices, and (4) sustainability, and (5) skeptics.

Farming Methods. Farming methods are one type of environmental concern. Angela lamented how at the traditional grocery store, farming methods are not evident on food product packaging. This invisibility contrasts with the food at the farmers' market, where you can ask the farmer how food was produced. Angela told me about how she and her partner Matt have become good friends with a local farmer who lives near them, and because of this friendship, they were able to see how this particular farmer produces food. She stated:

> I know not only do they have good methods, but also that I agree with their overall philosophy about how plants and land and animals should be treated. Because you can do organic and it can be awful—horrible, destructive monoculture. I can buy organic stuff at the supermarket, but I don't know the philosophy of the company or businesses.

Angela's added critique of monoculture reflects a counter-argument in favor of biodiversity, which arguably preserves species of vegetation and helps mitigate catastrophes like the Irish potato famine of the mid-19 century, which occurred due to an overdependence on a single crop (Pollan, 2001). Angela and Matt have found farmers at the market who meet their farming ideals. Matt recounted how one day a farmer named Jon asked them if they had read *One Straw Revolution* by Masanobu Fukoka. Matt said:

> Angela really likes the book. It's like this whole philosophical treatise on farming, basically. And so Jon mentioned this, and Angela was like "Yes" [enthusiastically], and it shows us that he has—maybe he doesn't live up to it, but an ideal.

Angela chimed in: "Basic ecological principles that like...and that's what, at least with Jon that we share." Angela and Matt are willing to pay more for food because they believe that what they are buying at the market represents the true cost of food. At the grocery store, they said, low prices obscure other costs like unethical practices, low wages, and a mismanagement of the environment. Similarly, Jaclyn believes that commercial farming practices are bad for the environment:

> I do think that one of the things that concern a lot of people with the big commercial farm is the use of either chemicals or water management practices or things that may overall be harmful to the environment. If your bottom line is all that you look at, then you try to produce the absolutely cheapest product you can and you can't always put concerns like that ahead of coming up with the lowest

price at the grocery store. These people [local farmers] are aware that their prices are not as low as the grocery store; they have to show you where they have additional value, and that additional value is in their farming practices, in their customer service. They have to give you a reason why you should pay more for their product.

For Jaclyn, farmers' market food has a higher value because these farmers prioritize food quality and the environment.

Pollution. The second sub-theme is pollution. As Jaclyn alluded to, the tension between pollution, pesticide use, and the potential solution in organic farming is important. Matt and Angela complained that pollution, in addition to other hidden costs of food, is often not accounted for with conventional agriculture. Matt commented:

> Economic externalities like causing pollution that you're going to have to pay for in the future, but you don't have to pay for emitting it in the process of production now. So it's not reflected—the future cost isn't reflected.

Angela added: "There's the environmental cost and the exploitation of the environment with the massive use of chemicals and destructive practices and things like that." Matt and Angela shop at the farmers' market in part because they want to avoid these environmentally destructive practices. Similarly, Lisa said:

> Because I'm a hazardous waste person, I see the environmental costs of producing pesticides and using pesticides, and I just don't think it's good for the planet and I don't think it's good for public health.

Lisa recently visited a company that has rights to "land-deploy" herbicides and pesticides. When I asked what this meant, Lisa clarified:

> Just spray them wherever they want in any quantity. I don't think most people know that. I was like, "Wow, really? Y'all can do that?" And I called the Department of Ag, and they were like, "Yeah, pretty much." It's kind of scarily deregulated.

Lisa mentioned that environmental protection was one important common value that she shared with farmers' market farmers. Pesticide use and pollution are significant concerns, and the potential to mitigate these practices via ethical organic farming is why many support the market. Not all farms at the market are organic, but most are. One vendor cleverly put up a sign saying "uncertified organic." Nevertheless, patrons can inquire about pesticide use and farming inputs.

Unethical practices. The third sub-theme deals with ethics. Not only are people concerned with environmental practices that are destructive, they are concerned with human and animal mistreatment surrounding food production. Jennifer shops at the farmers' market and grows her own food for some of these reasons. She uses her dollars and consumption patterns to voice her concern about the environment and agriculture. Jennifer said: "For me, it's a conscious statement as to how I want things to be. I believe in dollar bills as little ballots, and you're placing votes. I think it's the root of everything. Especially meat." Jennifer has qualms with current food production but hopes it can change. "I have problems with the food industry," she said. "Basically I'm just trying to do what I think is right so that in the future maybe this can become the norm." Jennifer believes that we need to consume wisely, think of future generations, and conserve our natural resources. When I asked what motivated her to support the farmers' market, she said:

> There are just horrible human, animal, environmental violations that are happening, specifically with the modern culture of corn. It's incredibly important to preserve our agricultural resources, and the way that we're doing so is—it's rapidly destroying our environment and our future, really.

Jennifer displayed erudite knowledge of environmental and food production practices. Similarly, Natalie, a sheep farmer who sells wool and lamb, advocates environmentally friendly farming and embodies that belief through her operation. Natalie explained:

> We are producing meat in a more expensive way because we're not doing the industrial thing. We're not feeding ammoniated straw to the cattle to fatten them, [nor] chemicals, added growth hormones. We're there as meat vendors because we're raising a few animals in an environmentally conscious way—in a humane way. That costs more.

Both Jennifer and Natalie make their performative stance in different ways as a result of their differing circumstances. Jennifer's embodied performance takes the form of careful consumption—"dollars as votes," and Natalie's expression means raising animals at a more costly price but which is more ethical and environmentally sensitive.

Sustainability. Sustainability is the fourth environmental sub-theme and overlaps with the previous three. According to patrons, sustainability is defined as progress, low environmental impact, and successful resource management. Carlo stated:

> The farmers' market is going to help make us better, and that by going to it, I'm helping contribute to that process. It's just part of progress. I don't know, I guess moral is definitely part of it, but definitely health-wise and environmentally.

For Carlo, simply going to the market is a participation in a counter-narrative. He further clarifies: "Reducing industry is something that occurs.... You're just reducing your impact maybe? At least your negative impact." When I asked Carlo if his motivations for going to the farmers' market were political, he said:

> I think subconsciously that there is an assumption that it is better for the environment, even though better for the environment is sort of abstract and there's no really real way of saying "Yes, the farmers' market does reduce your carbon footprint." And so I feel like those things are definitely in the dialogue when you're talking to people at the farmers' market. Now, do they occur every time that you buy something? That there's a little light that goes off in your head that says: "You just got green points!" I don't get that, but I do think that those are questions and buzzwords that are in a dialogue when you're trying to get people to go.

Carlo's statements reflect how one of the narratives of the market taps into this environmental "progress." Carlo mentioned that going green was a "buzzword" in the discursive market scene. Chris also felt like environmentalism was present at the market, but he thought it was an *unspoken* part of the market ethos. He said:

> There is probably a shared value with environmentalism. It's a leap, but I think it's there. No one talks about environment; we're not all there talking about pollution, but here we are, all walking around in an outdoor market buying locally grown food that is probably organic.

It is not surprising that there is an underlying, unspoken connection to social movements that encourage sustainability and environmental protection. Sustainability also means successfully managing renewable resources. Helen once worked in ocean conservation and has a specialized knowledge of fish farming. She stated:

> I come from a tiny town—it's huge for fish farming. But it's such a scourge on the environment. I think fish farms, if they were done better, in closed pens on land or something—they're necessary because we're completely eating our oceans. If we continue our fishing practices the way we are, we won't have any fish in 30 years.

Helen convinced me that she knew what she was talking about when she explained how the world's ocean fisheries are threatened today with over-harvesting. Unlike some who might be opposed to fish farms, Helen realizes that they may be a practical necessity if we are to preserve the environment. Bruner and Meek (2011), in their discussion of seafood politics, point out that some reports do show that if fishing practices are not regulated properly the world may run out of seafood by the year 2050. Helen's concerns are valid, and the farmers' market, while it does not sell fish, connects to the larger sustainability narrative.

Others critique mainstream food systems' lack of sustainable practices. Jaclyn cited how Kansans import soybeans from China when they are grown locally. She rhetorically asked: "How can that possibly be a good idea? How can it, number one, be cheaper? Number two—that is an awful lot to be putting into the environment from the simple shipping. So it's totally unnecessary."

Sustainability also takes specific forms like preserving pristine farmland. Natalie believes that her customers support not merely her business but also the preservation of the good farmland in the north part of town that is threatened by development. "They're supporting the use of sustainable technologies; they know that I don't use a lot of fossil fuels. I don't use a lot of irrigation. I'm not buying a lot of inputs for the farm." Sheryl, too, mentioned the preservation of soils:

> We need to take care of it and not put in parking lots and understand that we've got some class A1 soils north of Lawrence. These are really a precious resource, and somebody [developer] wanted to put an industrial park on top of them.

Fearful of ignorant development, Sheryl expressed her disdain for developers who threaten to damage this resource.

Skeptics. Finally, many see the market as a way to escape environmental problems. On the other hand, a few questioned whether or not local farms were actually ecologically advantageous, and their voices are worth noting. Chris stated:

> Small-scale farming isn't necessarily green, from what I've read. It's an interesting study, if you read *Mother Jones*, they're going to say, "Hey, this is the only way there is to sustain the planet." But if you read the *New York Times* you'll see economists and smart people who have done the math and say, "Look, this level of farming is never going to support the population." And the guys over here in *Mother Jones* magazine are kidding themselves if they think we're going to feed the nation by growing stuff this way.

Helen, too, was skeptical about whether small farming practices were actually going to improve food system problems. She said:

> I'm such a cynic about all of that stuff because, like the organic salmon. I think we all have to be a lot more critical, and it's never just, "Oh buy local and everything's okay" or "Oh, buy organic." I think it's a lot more complex than that. To say that by shopping at a farmer's market you're being more green or more environmentally friendly—that might not actually be the case. And it's really specific, even down to potentially the specific vegetables that are produced and how they're produced and the amount of energy that goes into making them.

Helen suggested visiting the farm where you buy from to see how they operate. She was worried about farming practices, and I asked her to summarize these concerns. She said: "Sometimes depending on the farming practices there might be more greenhouse gases consumed locally than if they're done further away and shipped." Helen went on to question whether local production is actually better for the environment. She suggested that large-scale crops may in fact be more efficient and less environmentally harmful.

These narratives reflect a concern for the environment and show how participants are performing activism. Most of the time the discourse surrounding the environment is implicit. As Chris said, environmentalism is a fundamental but unspoken rationale for going. More explicit environmental messages are reflected in market pamphlets, websites, and vendor signage. There are also skeptics like Helen and Chris who support the market but maintain a critical eye toward the assumption that local is more sustainable.

Conclusion

The farmers' market allows a space for alternative voices to be heard that critique problems with industrial agriculture. Patrons enact this voice by supporting farmers monetarily, forming emotionally supportive friendships with farmers, and by placing their bodies in the market. Unethical farming practices, pollution, unethical treatment of animals and laborers, and the lack of sustainability encompass some of these fears. Environmental concern represents only one of many motivations for supporting farmers' markets, but it is an important underlying rationale for this type of food system.

Enacting support in a public venue heightens the performative element by adding a layer of reflexivity. Unlike a CSA, farmers' market exchanges occur in a public venue that creates a type of spectacle inherent to performances (Bell, 1997; 2004). This particular example corroborates the claim that American citizens use their consumption in political ways (Pezzullo, 2011; Zukin et al., 2006). Some, like Jennifer, consider this patronage a "vote" for better agriculture. Yet Jennifer's vote is not an anonymous casting of ballots. Rather, she and other customers place their bodies in the market space, purchase products face to face, and publicly perform their support. Performative patronage is one manifestation of voice—giving an account of one's values. The farmers' market context fits the four elements of performance defined previously: it has an audience, reflects community values through cultural performance, emphasizes embodiment, and occurs in a public setting.

The performance of values at the farmers' market is reaching audiences. Couldry (2010) argues that listening and recognition are important steps in countering the neoliberal rationality of exchange because these acts restore reciprocity. Customers at the market have a plethora of audiences for which to perform. Other citizen-patrons, farmers, the self, and city officials listen to and validate this environmental message. Interacting with other consumers generates solidarity, and dialogue with farmers also validates environmental values. The audience of self is important, too. Performance theorists have long recognized the way people observe their own behavior in a double consciousness way, comparing it to past or other performances, in order to maintain their identity, face, or front (Carlson, 2004; Goffman, 1967; 2004). At the same time, Couldry (2010) recognizes that the institutions have the power to "undermine" voice by retaining practices that deny voice (p. 10). Recently, the Lawrence farmers' market management rallied loyal customers and vigorously petitioned the city commissioners to allow the market access to a particular location for their weekday market, and the city allowed this despite resistance from nearby businesses (Riling, 2013). This indicates farmers' market voices can and do influence *local* politicians. However, it is unclear to what extent corporate agriculture recognizes or listens to market voices. Yet I do not believe support or denial from corporations negates the affirmation of voice that farmers, other citizen-customers, and city officials provide in recognition of the importance of alternate agriculture.

Second, the farmers' market also represents a cultural performance where communal values are displayed in a reflexive way. While not everyone attends the market for environmental reasons, some do. The public venue of the farmers' market contextualizes market patronage within a communal gathering place, a public square experience that transends private consumption. Market participation ties into Watts' (2001) notion that voice represents "the sound of specific experiential encounters in civic life" (p. 185) because market patrons express and deal with fear and concern about the environment in part by supporting this alternate form of agriculture. Not everyone that I interviewed had read books on environmentalism or talked about it reflectively, but many did. The market provides an opportunity to enact philosophical and ethical values.

Third, market participation also emphasizes embodiment. Patrons walk around, shop, compare foods, and purchase their goods in a public venue. For many, this ritual performance occurs every Saturday morning with family, pets, and coffee in hand. Market patrons promenade around the market performing values, similar to how Italians performed modernity in Del Negro's (2004) study. Using one's body in public space persuasively enacts voice. Purchasing organic food at the grocery or subscribing to a CSA exemplifies more private support for alternate agriculture. Market consumption, by contrast, is public spectacle.

Performance is deeply communicative. It creates self and collective definition. Carlo, for example, once ate conventional food out of convenience, but through the market he redefined his habits and ideology. By continuing to support the market, he also supports his new identity as an environmentally conscious individual. The DLFM organization and the people I interviewed are performing a collective message and identity of alternative agriculture, even though individual expression may differ.

The success of farmers' markets in places like Lawrence is partly due to a deep mistrust of monolithic food corporations that are perceived to prioritize profits over ethics. Though still having only a small percentage of the market share, farmers' markets represent food activism in action—a voicing of concern by consumption, participation, and performance. The collective and individual performances at the DLFM suggest that local food systems are one way to fix the problem. Farmers, other citizen-patrons, the self, and city officials validate this voice as important. At the current juncture, more studies are needed to assess how corporate audiences are responding to patron voices and farmers' market national growth. Additional studies with diverse methodologies might examine what role location and context play in affecting patron values and attitudes.

References

Alkon, A. (2008a). From value to values: Sustainable consumption at farmers markets. *Agriculture and Human Values, 25*, 487–498.

Alkon, A. (2008b). Paradise or pavement: The social constructions of the environment in two urban farmers' markets and their implications for environmental justice and sustainability. *Local Environment, 13*, 271–289.

Andreatta, S., & Wickliffe, W. (2002). Managing farmer and consumer expectations: A study of a North Carolina farmers market. *Human Organization, 61*, 167–176.

Baxter, L. A., & Braithwaite, D. O. (2002). Performing marriage: Marriage renewal rituals as cultural performance. *Southern Communication Journal, 67*, 94–109.

Bell, C. (1997). *Ritual: Perspectives and Dimensions*. Oxford: Oxford University Press.

Bell, C. (2004). "Performance" and other analogies. In H. Bial (Ed.), *The Performance Studies Reader* (pp. 98–106). London: Routledge.

Bourdieu, P. (1984). *Distinction: A Social Critique of the Judgement of Taste*. Cambridge MA: Harvard Univ Press.

Braun, V., & Clarke, V. (2006). Using thematic analysis in psychology. *Qualitative Research in Psychology, 3*, 77–101.

Brown, C., & Miller, S. (2008). The impacts of local markets: A review of research on farmers markets and community supported agriculture (CSA). *American Journal of Agricultural Economics, 90*, 1296–1302.

Bruner, M. S., & Meek, J. D. (2011). A critical crisis rhetoric of seafood. In J. M. Cramer, C. P. Greene & L. M. Walters (Eds.), *Food as Communication: Communication as Food* (pp. 271–295). New York: Peter Lang.

Carlson, M. (2004). What is performance? In H. Bial (Ed.), *The Performance Studies Reader* (pp. 70–75). New York: Routledge.

Charmaz, K. (2006). *Constructing Grounded Theory: A Practical Guide through Qualitative Analysis*. London: Sage.
Couldry, N. (2010). *Why Voice Matters: Culture and Politics after Neoliberalism*. London: Sage Publications Ltd.
Del Negro, G. (2004). *The Passeggiata and Popular Culture in an Italian Town: Folklore and the Performance of Modernity*. Montréal, QC: McGill-Queen's University Press.
Downtown Lawrence Farmers' Market. (n.d.). Retrieved 1 May 2012, from http://www.lawrencefarmersmarket.com/
Emerson, R. M., Fretz, R. I., & Shaw, L. L. (1995). *Writing Ethnographic Fieldnotes*. Chicago: University of Chicago Press.
Faber, A. (2004). Saint Orlan: Ritual as violent spectacle and cultural criticism. In H. Bial (Ed.), *The Performance Studies Reader* (pp. 118–124). New York: Routledge.
Farmers market search. (2013). Retrieved 25 January 2013, from http://search.ams.usda.gov/farmersmarkets/
Feagan, R. B., & Morris, D. (2009). Consumer quest for embeddedness: A case study of the Brantford farmers' market. *International Journal of Consumer Studies, 33*, 235–243.
Giddens, A. (1979). *Central Problems in Social Theory: Action, Structure, and Contradiction in Social Analysis*.
Giddens, A. (1984). *The Constitution of Society: Outline of the Theory of Structuration*. Berkeley: University of California Press.
Goffman, E. (1967). *Interaction Ritual: Essays on Face-to-Face Behavior*. Garden City, NY: Doubleday.
Goffman, E. (2004). Performances: Belief is the part one is playing. In H. Bial (Ed.), *The Performance Studies Reader* (pp. 61–69). New York: Routledge.
Gowin, J. (2009). Meet your maker. *Psychology Today, 42*, 51–52.
Kenner, R. (Writer). (2008). *Food, Inc*. Magnolia.
Kloppenburg, J., Lezberg, S., De Master, K., Stevenson, G. W., & Hendrickson, J. (2000). Tasting food, tasting sustainability: Defining the attributes of an alternative food system with competent, ordinary people. *Human Organization, 59*, 177–186.
Lyson, T. A., Gillespie Jr, G. W., & Hilchey, D. (1995). Farmers' markets and the local community: Bridging the formal and informal economy. *American Journal of Alternative Agriculture, 10*, 108–112.
MacAloon, J. J. (1984). Introduction: cultural performances, culture theory. *Rite, Drama, Festival, Spectacle: Rehearsals toward a Theory of Cultural Performance* (pp. 1–15). Philadelphia: Institute for the Study of Human Issues.
McGrath, M. A., Sherry, J. F., & Heisley, D. D. (1993). An ethnographic study of an urban periodic marketplace: Lessons from the Midville farmers' market. *Journal of Retailing, 69*, 280–319.
Opel, A., Johnston, J., & Wilk, R. (2010). Food, culture, and the environment: Communicating about what we eat. *Environmental Communication, 4*, 251–254.
Opler, M. E. (1945). Themes as dynamic forces in culture. *American Journal of Sociology, 51*, 198–206.
Owen, W. F. (1984). Interpretive themes in relational communication. *Quarterly Journal of Speech, 70*, 274–287.

Pezzullo, P. C. (2011). Contextualizing boycotts and buycotts: The impure politics of consumer-based advocacy in an age of global ecological crises. *Communication and Critical/Cultural Studies, 8*, 124–145.
Pollan, M. (2001). *The Botany of Desire: A Plant's-Eye View of the World*. New York: Random House Trade Paperbacks.
Pollan, M. (2007). *The Omnivore's Dilemma: A Natural History of Four Meals*. New York: Penguin Press.
Pollan, M. (2008). *In Defense of Food: An Eater's Manifesto*. New York: Penguin Press.
Preston, J. (2012). Chick-fil-a draws huge crowds for appreciation day. Retrieved 19 August 2012, from http://thelede.blogs.nytimes.com/2012/08/01/chick-fil-a-draws-huge-crowds-for-appreciation-day/
Retzinger, J. P. (2008). Pizza as praxis: Bridging nature and culture. *Environmental Communication, 2*, 246–255.
Riling, M. E. (2013, March 12). Your turn: Market move raises concerns. Retrieved 14 March 2013, from http://www2.ljworld.com/news/2013/mar/12/your-turn-market-move-raises-concerns/?opinion
Robinson, J. M., & Hartenfeld, J. (2007). *The Farmers' Market Book: Growing Food, Cultivating Community*. Bloomington, IN: Indiana University Press.
Ryan, G. W., & Bernard, H. R. (2003). Techniques to identify themes. *Field Methods, 15*, 85–109.
Salbu, S. R., from (2012). Let Chick-fil-a fly free. Retrieved 12 August 2012, from http://www.nytimes.com/2012/08/02/opinion/let-chick-fil-a-fly-free.html
Spurlock, C. (2009). Performing and sustaining (agri)culture and place: The cultivation of environmental subjectivity on the Piedmont farm tour. *Text and Performance Quarterly, 29*, 5–21.
Stull, D. D., & Broadway, M. J. (2004). *Slaughterhouse Blues: The Meat and Poultry Industry in North America* (Vol. 1). Belmont, CA: Wadsworth.
Todd, A. M. (2011). Eating the view: Environmental aesthetics, national identity, and food activism. In J. M. Cramer, C. P. Greene & L. M. Walters (Eds.), *Food as Communication: Communication as Food* (pp. 297–315). New York: Peter Lang.
Trobe, H. L. (2001). Farmers' markets: Consuming local rural produce. *International Journal of Consumer Studies, 25*, 181–192.
Warren, C., & Karner, T. (2010). *Discovering Qualitative Methods: Field Research, Interviews, and Analysis*. New York: Oxford University Press.
Watts, E. K. (2001). "Voice" and "voicelessness" in rhetorical studies. *Quarterly Journal of Speech, 87*, 179–196.
Watts, E. K. (2012). *Hearing the Hurt*. Tuscaloosa: University of Alabama Press.
Winne, M. (2008). *Closing the Food Gap: Resetting the Table in the Land of Plenty*. Boston: Beacon Press.
Zukin, C., Keeter, S., Molly, A., Jenkins, K., & Carpini, M. X. D. (2006). *A New Engagement?: Political Participation, Civic Life, and the Changing American Citizen*. New York: Oxford University Press.

8
Response Essay: Thinking through Issues of Voice and Consumption

Laura Lindenfeld

In the United States, as in many other post-industrial nations, what we consume functions as an important part of how we define ourselves collectively as citizens. From this vantage point, citizenship is clearly more than just voting behavior. It encompasses our daily rituals and practices, which includes our consumption of goods and services (Garcia Canclini, 2001). Culture and economics co-define citizenship along with the political in complex ways that invite us to foreground the mundane choices we make in everyday life and to consider how consumption, identity, and voice intersect. Defining ourselves by our consumption can create the illusion that we are expressing our own taste and personality through our consumption, when indeed the options available to us are often pre-determined and pre-packaged by marketing and production constraints. What role might the concept of voice play in helping us to tease out the relationship between dominant discourses and patterns that limit our choices and individual agency? The essays in this section turn to this question and ask where voice and consumption intersect and how we can conceptualize a more nuanced understanding between the material and political dimensions of consumption and the relationship between consumer choices and agency.

If we seek to understand how people give voice to and listen to the voices of the environment—or render these voices silent—then we must create deeper insights into the relationship between consumption and voice. As the two essays in this section illustrate, this understanding requires critical listening that parses out the complex, often contradictory intersections among identity, politics, consumption, and voice.

Leah Sprain's and Benjamin Garner's essays do precisely this kind of careful, nuanced work. Together, they grapple with the relationship between voice and consumption in novel ways. Through close analyses of many different voices, they consider how communication about the environment—in both cases communication about and through organic food—navigates politics, consumption, and voice. As Sprain concludes in her study, consumption functions as a form of speech. Organic consumption "talks." Similarly, Garner concludes that farmers' markets enable a certain group of consumers to place their bodies within a public setting in such a way that they can communicate messages about the environment to diverse populations. Here, individuals function collectively as a mouthpiece for the environment in ways that attempt to privilege voices otherwise repressed. For both authors, voice is defined by process and by the tensions between conflicting value systems and socio-political hierarchies in which it emerges. Voice is performative and it is rhetorical, actively shaping and defining how we experience each other and the world around us (Couldry, 2010; Watts, 2001). Both essays remind us of the centrally important function of communication as constitutive, as a force that shapes our world and our experiences of each other. Expressing voice on behalf of the environment represents a form of power as well, as voice can take on the form of "talking back" against dominant structures that often seek to render the environment voiceless. In different but complementary ways, both essays investigate and illustrate the complexity of this relationship.

Sprain's chapter asks the reader to think about how individuals articulate food as political in online fora, while analyzing discourses across a range of media, including environmental and activism websites, magazines, books, and promotional brochures. Her thorough analysis of different articulations of voice reminds us how important it is to bring the tools of rhetorical criticism and the insights of rhetorical theory to this conversation. Her essay looks at voice primarily through a political lens, a particularly important vantage point, since consumption and politics are conceptually intertwined. Following two interwoven threads of discourse—that of consumers/individuals and that of various media—enables her to identify key themes that emerge as dominant arguments supporting organic consumption, particularly those coming from advocacy organizations. We get a sense of the broader landscape in which this conversation is currently taking place. Sprain looks effectively to theories of voice to situate the "political potential of organic consumption within this shifting landscape" (this volume, p. 142).

Garner's chapter also explores politics, food, activism, and voice in ways that complement Sprain's analysis. Garner argues that farmers' markets represent a form of political voice that articulates itself in opposition to conventional agriculture and megacorporations that suppress the voice of the environment. Through a performance analysis of farmers' market consumers, Garner explores how people can physically place themselves in a public to communicate environmental messages. The preponderance of environmental concerns within this discursive context mobilizes Garner's performative approach, and he defines voice as the process and value that emerge through narrative analysis. Voice serves here as critique of corporate agriculture and a physically embodied performance that underscores the values embraced by the farmers' market.

These two essays provide important insights that advance our understanding of voice and consumption, and both scholars pushed me to think about consumption and voice in new ways. Collectively, and individually, they offer important insights into future research. In particular, my response seeks to articulate four key intersections that emerge across these two essays. I outline these briefly here and situate them vis-à-vis broader scholarly conversations about consumption. As I lay out these key intersections, I consider what kinds of questions these two essays collectively raise that adds to the current knowledge base, how they push our understanding of voice and consumption into important new frameworks, and how this work can help to inform future inquiry and practice.

Intersection 1: Food, consumption, and voice

It is no surprise that essays on consumption and voice choose food as an object of study. As food scholar Sidney Mintz writes, "food and eating afford us a remarkable arena in which to watch how the human species invests a basic activity with social meaning—indeed, with so much meaning that the activity itself can almost be lost sight of" (Mintz, 1996, p. 7). Food seems an obvious locus for this conversation. It represents an everyday object of consumption that we *must* consume to exist. Food literally becomes part of our bodies; it is essential to creating and preserving the bodies that enable us to even have a voice. Food is also a part of everyday life that can otherwise easily be swept under the carpet as a site of political tension, and yet it is profoundly political. Its seemingly everyday nature enables us to think through the many aspects of daily life where the cultural, social, and political realms intersect. Food, in my eyes, is the ideal locus for thinking about consumption and voice. Food masks its political, economic, and cultural contexts and renders it easy for many people to forget that food is about so much more than eating.

These two essays offer a refreshing set of insights into how food articulates political power, and where the limitations of that power might be. They identify contexts where individuals and groups refuse to accept the seemingly easy reduction of food to the realm of the apolitical. Indeed, they expose this very tendency and focus instead on sites where food explicitly becomes a vehicle for expressing voice about the environment. Sprain and Garner help us to think about how fundamentally linked our societal sense of belonging, empowerment, and enfranchisement are across different kinds of consumption through the lens of food.

A significant body of work on food production, consumption, and politics has emerged over the past decade. Eric Schlosser's (2001) interrogation of fast food production, Michael Pollan's (2006) call to change the food system, and the work of many others have advanced the conversation about food in sophisticated ways so that many people know that food is not just about food. More recently, as a field communication has come to the proverbial food table and made important contributions to the conversation about food, power, identity, and communication (Cramer, Greene, & Walters, 2011; Frye & Bruner, 2012). What these two essays add to the literature is that they help us understand how organic food and sustainable approaches to food more specifically function as contexts for articulating voice. This adds to our understanding of the relationship between food and power, an area that has focused less on the environment within the research on food and communication. This perspective is particularly refreshing because many forms of food consumption depend on strategies that suppress voice rather than articulating it. Here, consuming and eating food becomes a form of deliberate, conscious articulation, and the politics and practices of our daily lives are laid bare.

Focusing on food, voice, and the environment represents an important step for future inquiry and practice. These two chapters demonstrate that food and voice can help to connect environmental issues to broader contexts, and food functions as a fertile basis from which to circle back to other broader issues.

Intersection 2: Producing depth through ethnographic approaches

Similar to their focus on food, both essays use ethnographic methods to approach the issue of voice and consumption. This synergy is compelling and indicative of where the conversation about voice, consumption, and the environment currently stands. Inductive ethnographic methods enable both authors to let the voices of individuals and groups emerge (Middleton, Senda-Cook, & Endres, 2011). I find this intersection particularly noteworthy, as it enables us to focus on how individuals and groups are constructing voice in real-world settings. Sprain approaches her study with an ethnographic analysis that integrates rhetorical analysis of promotional materials with a discourse analysis of Usenet discussion groups. Garner's interviews are performance-based in their approach to how individuals place their bodies in public contexts to communicate about food in environmental terms. Sprain's substantive data set—looking to online discussions to track discourse, while simultaneously looking at documents—represents a strategy that affords us a broader sense of how some people are giving voice to organic consumption, whereas Garner's analysis of farmers' markets lets us hear the voices of a particular group of consumers and their perspectives.

Both represent productive uses of ethnography that integrate different kinds of analysis with participant observation, one virtual and one in the physically experienced space of the farmers' market. Reading across these two essays, I found myself imagining the overlap in these different communities of study, farmers' markets, and activist websites. I envisioned these voices as having some overlap, but also some differences, and the ability to tease out these differences and similarities across larger groups will be important to future research in this area. Ethnographic approaches enable us to look at the picture of these two respective contexts from a slightly different vantage point, and they utilize rich analysis of dialogue to initiate this scholarly conversation in ways that produce analytical depth that other frameworks do not. Both essays are concerned with drawing attention to details that enable us to gain deeper, crucial insights into to understanding how food and voice work in these particular respective settings.

The choice of ethnography reveals something to me about where the broader scholarly conversation about voice and consumption currently stands in the field of communication: we need to attend to the voices of particular individuals and groups. This recognition led me to consider where compelling future questions might emerge. How might scholars build on this detailed ethnographic understanding to develop broader, empirically based understandings of how organic consumption and voice intersect? How might we maintain the depth of understanding that emic perspectives enable us to experience while assessing the way discourses travel across broader contexts and circles? Starting this conversation with ethnography is an important move. In the future, I imagine that we will need more ethnographic work, but we will also need to broaden the toolbox and try to figure out how different perspectives and methodological approaches can help to broaden the questions and conceptualize answers that build on the deep understanding that ethnography enables us to answer. Some of these questions will require interdisciplinary collaborations to create appropriate answers. One could envision researchers from a variety of fields coming together to shape questions and approaches that let us build on the insights gained through the work of researchers like Sprain and Garner.

Intersection 3: Foregrounding the political

The foregrounding of political issues creates an obvious synergy between these two essays. Both are explicit about the relationship of food, the environment, voice, and the political. Both essays configure voice as an expression of explicit politics, and both grapple with this configuration. This shift in seeing voice and food as a deliberate, conscious form of politics is refreshing and inspiring. In thinking about the history of food studies, much scholarship has considered the unconscious politics associated with food, especially feminist scholarship on food and voice as these relate to eating disorders (Bordo, 2004; Brown & Jasper, 1993; Fallon, Katzman, & Wooley, 1994). Other perspectives have explored the politics of food, sometimes with a focus on voice, but without necessarily attending to issues of communication, power, and voice (Belasco, 2007; Nestle, 2002; Watson & Caldwell, 2005). These two essays build on the work of scholars like Belasco, Nestle, Watson, and Caldwell to situate food as a public form of expressing voice that is specifically political. Garner attends to the conspicuous, physical issue of "being there," that is of performing voice at the farmers' market, and considers how storytelling becomes a political act, wherein the farmers' market functions as political stage. In Sprain's essay, one of the key themes of voice is that of "dollars as votes" and "voting with one's fork," which reaffirms the configuration of individual behaviors as political acts.

The concept of voting is key to both texts. As scholars like Canclini and Miller remind us, consumption and politics are inextricably linked (García Canclini, 2001; Miller, 2007). Food has been recognized as an important site to explore the contestation of cultural citizenship, the kind of citizenship that appears in everyday life in the consumption of goods and services, including leisure activities and entertainment (Miller, 2007). This concept of cultural citizenship stresses that culture is central to our participation in society, and it reminds us that citizenship is much more than voting (political citizenship) or purchasing goods (economic citizenship) (García Canclini, 2001).

Intersection 4: Multiple perspectives—it's complex!

While there are many intersections that emerge across these two essays—intersections with each other and with the broader tapestry of scholarship and discourse on food, voice, and the environment—a final one that deserves some explicit attention in the context of this book is the theme of complexity. Both Sprain and Garner understand that they have stepped into complex, ever-changing territory, where the discourse itself shapes not only the conversation, but also the impacts that these conversations have on the environment. Both recognize that we are looking at multiple sides of a complex, evolving issue, with Garner articulating the possibility of demonstrating the possibility of resistance to environmental degradation through physical participation in farmers' markets and Sprain reminding us that this resistance can easily be co-opted. Together, these essays represent a productive critical tension that many of us grapple with: how can individual expressions of resistance counter dominant discourses and practices that easily co-opt and undermine them? What are the possibilities of voice? What are the limitations, and how can some voices drown out others? Garner's enthusiasm and empowering analysis inspire hope; Sprain's reminder of how neoliberalism works to elevate certain types of voice at the expense of others. As communication researchers, we must attend precisely to this tension, and we need both sides of the coin: the inspirational and the critical. Voice is powerful, and it can and does change relationships and practices, but as Sprain reminds us, responsible relationships to food are easily undermined if we only let markets listen to organic consumers' voices. As Sprain concludes, there are indeed ways in which groups, like the Usenet group she analyzes, have developed codes for "understanding consumption as a means of strategic action that speaks and makes demands for a particular type of food system" (this volume, p. 143).

Looking forward

These essays add new, important insights into a number of ongoing scholarly conversations, and they represent important tensions, important intersections, and important differences. Both of these pieces strive for depth. A future challenge will be to develop strategies that enable us to maintain this richness, yet still understand the breadth of voice and consumption. Here, the conversation has focused on the obvious thematic choice of food. How might other conversations about different topics differ, and where can issues of voice related to other topics—like water, or climate change, or endangered species—inform our understanding of voice and food?

Personally, I always struggle with the question of how we move what we know into what we can do. Increasingly, fields like sustainability science are issuing a call to us as researchers to become directly engaged in solving problems (Clark & Dickson, 2003; Clark & Holliday, 2006; Hart & Calhoun, 2010; Kates, 2011; Komiyama & Takeuchi, 2006; Matson, 2009). My own work has increasingly turned to identifying how we can make communication research more engaged in oriented work (Lindenfeld, Hall, McGreavy, Silka, & Hart, 2012). Communication perspectives have the sophisticated capacity to tease out the relationship between issues of voice and power, and I believe it is part of our obligation as environmental communication scholars in particular to align our research with the pressing problems society faces (Cox, 2007). These two essays on food, voice, and power have important contributions to make within our field and outside of it, both to other researchers and to different kinds of decision-makers and communities.

The question then becomes one of how can we become engaged directly with communities and have our work enter more directly into decision-making processes. These essays help us understand some of the current ways that people are articulating voice around the environment and food, but how do we create voice for communication researchers within these contexts to help articulate change? Food and voice represent a particularly relevant context and a rich space for advancing this aim. Direct partnerships with farmers' markets, sustainable food networks, and K-12 school efforts to integrate more organic, local foods, for example, might offer some productive opportunities for partnerships that can lead to real-world solutions, where communication research plays a central role. The central question here is: how can our engagement as researchers help us advance the important efforts of different groups while countering challenges? Sprain emphasizes potential for political voice when organic consumption is viewed as social movement behavior, but this potential has not yet been broadly realized. I believe there are numerous ways in which our direct engagement as researchers can involve us in processes that integrate our knowledge and efforts with that of communities.

References

Belasco, W. J. (2007). *Appetite for Change: How the Counterculture Took on the Food Industry* (2nd updated ed.). Ithaca: Cornell University Press.

Bordo, S. (2004). *Unbearable Weight: Feminism, Western Culture, and the Body* (10th anniversary ed.). Berkeley, Calif.: University of California Press.

Brown, C., & Jasper, K. (1993). *Consuming Passions: Feminist Approaches to Weight Preoccupation and Eating Disorders.* Toronto: Second Story Press.

Clark, W. C., & Dickson, N. M. (2003). Sustainability science: The emerging research program. *Proceedings of the National Academy of Sciences of the United States of America, 100(14),* 8059.

Clark, W. C., & Holliday, L. (2006). Linking knowledge with action for sustainable development: The role of program management—summary of a workshop. *Roundtable on Science and Technology for Sustainability.* Washington, DC: National Research Council. Retrieved from http://www.nap.edu/catalog.php?record_id=11652

Couldry, N. (2010). *Why Voice Matters: Culture and Politics after Neoliberalism.* Los Angeles, London: SAGE.

Cox, R. (2007). Nature's "Crisis Disciplines": Does Environmental Communication have an ethical duty? *Environmental Communication, 1(1),* 5–20.

Cramer, J. M., Greene, C. P., & Walters, L. (2011). *Food as Communication: Communication as Food.* New York: Peter Lang.

Fallon, P., Katzman, M., & Wooley, S. (1994). *Feminist Perspectives on Eating Disorders.* New York: Guilford Press.

Frye, J., & Bruner, M. S. (2012). *The Rhetoric of Food: Discourse, Materiality, and Power.* New York: Routledge.

García Canclini, N. (2001). *Consumers and Citizens: Globalization and Multicultural Conflicts.* Minneapolis, London: University of Minnesota Press.

Hart, D. D., & Calhoun, A. J. K. (2010). Rethinking the role of ecological research in the sustainable management of freshwater ecosystems. *Freshwater Biology, 55,* 258–269. doi: 10.1111/j.1365-2427.2009.02370.x

Kates, R. W. (2011). What kind of a science is sustainability science? *Proceedings of the National Academy of Sciences of the United States of America, 108(49),* 19449–19450. doi: 10.1073/pnas.1116097108

Komiyama, H.,& Takeuchi, K. (2006). Sustainability science: Building a new discipline. *Sustainability Science, 1(1),* 1–6. doi: 10.1007/s11625-006-0007-4

Lindenfeld, L. A., Hall, D. M., McGreavy, B., Silka, L., & Hart, D. (2012). Creating a place for Environmental Communication research in Sustainability Science. *Environmental Communication, 6(1),* 23–43. doi: 10.1080/17524032.2011.640702

Matson, P. (2009). The sustainability transition. *Issues in Science & Technology, 25(4),* 39–42.

Middleton, M. K., Senda-Cook, S., & Endres, D. (2011). Articulating rhetorical field methods: Challenges and tensions. *Western Journal of Communication, 75(4),* 386–406.

Miller, T. (2007). *Cultural Citizenship: Cosmopolitanism, Consumerism, and Television in a Neoliberal Age.* Philadelphia: Temple University Press.

Mintz, S. W. (1996). *Tasting Food, Tasting Freedom: Excursions into Eating, Culture, and the Past.* Boston: Beacon Press.

Nestle, M. (2002). *Food Politics: How the Food Industry Influences Nutrition and Health*. Berkeley: University of California Press.

Pollan, M. (2006). *The Omnivore's Dilemma: A Natural History of Four Meals*. New York: Penguin Press.

Schlosser, E. (2001). *Fast Food Nation: The Dark Side of the All-American Meal*. Boston: Houghton Mifflin.

Watson, J. L., & Caldwell, M. L. (2005). *The Cultural Politics of Food and Eating: A Reader*. Malden, MA: Blackwell Pub.

Watts, E. K. (2001). "Voice" and "voicelessness" in rhetorical studies. *Quarterly Journal of Speech, 87*, 179–196.

Section III
Listening to Nonhuman Voices

9
The Language That All Things Speak: Thoreau and the Voice of Nature

William Homestead

Henry David Thoreau was prescient, recognizing in the mid-19th century that our allegiance to industrialization was leading us along a problematic path. His cabin at Walden Pond was only one and a quarter miles from town, and Concord was already a mix of farms and burgeoning industry, a mere 16 miles from the expanding city of Boston. Two-thirds of New England, excluding Maine, was cleared land by 1837. Industrialization was creeping forward, displacing nature's web with an interconnected web of railroads. Concord was a pleasant place to explore, filled with pastures, meadows, swamps, and woodlands of six to ten acres each, but Thoreau wrote in his journal that it was impossible to walk in the woods during the day without hearing the chopping of an axe. And he could see and hear trains from his cabin, which provided a constant reminder of a world transformed by the dictates of humans (Richardson, 1986).

Today, we need not be prescient to realize that industrial narratives of unlimited growth (bigger), technology as savior (better), and linear progress (more of the same) are leading us to conduct a grand ecological experiment; we simply need to understand a basic principle of ecology, interdependency, and read headlines of oil spills, species extirpation, and the like. The ecosystem is too complex for us to predict what will happen, but we know the litany of eco-social ills that have already happened, and we know that there are numerous signs of what may happen. Yet, although we have heard the scientific warnings, and experienced the desecration and loss of beloved places, we still don't seem to be getting the message. The ecotheologian Thomas Berry (1988) argues that the natural world has ceased to flow into our physical and psychological being and, thus, we cannot properly listen and respond. This argument raises serious concerns, especially when we consider anthropologist and systems theorist Gregory Bateson's (1972) claim that the biosphere is best understood as a communicative system of information exchange. We face an ecological, and communicative, dilemma: the ecosystem is resoundingly dialogic—everything exists within relationships of sending and receiving, including nonverbal presence, exchange of energy, and the oxygen–carbon dioxide cycle—and yet we have created socio-cultural systems that are largely governed by monologue.

Monologic, or instrumental, attitudes and forms of knowledge have a function—the earth supplies resources and objective scientific studies provide insights, for example—but it is unresponsive and irresponsible to forget that nature is the source and sustenance of our very being. Thus, our penchant for monologue, for separating ourselves from and attempting to control nature (telling it what to do), needs to be contextualized and rethought within the possibilities of dialogic wisdom, or the learning that presents itself when we recognize the primacy of relationship and reciprocity, and, ultimately, a spiritual unity in material diversity. Thoreau (in Emerson & Thoreau, 1991) knew, intellectually and experientially, that we are "part and parcel" of the natural world, and that human development depends on our willingness to attend to calls from nonhuman others, whether animal, tree, plant, or insect, being continually confronted with the ethical question of how we should respond (p. 71). Or rather, we have "response-ability," but our monologic muting of nature causes us to miss out on way too much.

What I am calling dialogic wisdom entails a widening of possible modes of listening and learning that includes pre-rational, rational, and transrational experience. Said differently, upon consulting Thoreau's seminal texts (*Walden*, "Walking"), the "heart" of his journals, and several biographies and critical essays, I argue that Thoreau advocates for pre-rational bodily empathy and sensuous listening to the voices of nature; acute scientific observation and contemplation, inspired by aesthetic sensibility and poetic imagination, of the voices of nature; and transrational experience of the *Voice* of nature, apprehending a spiritual unity among material diversity, or transcendent Spirit within every material manifestation. All are necessary if we are to fully embody Martin Buber's (1970) claim that living means being addressed; all are necessary for an ethical responsiveness to ecocrisis.

The experience of unity, which I am calling "the voice of nature," should not be seen as creating a totality where diverse voices disappear or are devalued, nor should it be seen as a transcendentalist ideal separating humans from nature, but as a moment of increased clarity. In fact, nondual mystical experience, often discovered in steps via the senses, observation, and contemplation, provides a key insight, challenging our penchant for monologue governed by a damaging duality between mind and nature. This duality, or our ideological and psychological separation from nature (there is no ecological or spiritual separation), has given birth to the dominant narratives of bigger, better, and more. And we live the stories we tell, creating educational, judicial, religious, and political institutions that structurally deny and unconsciously and habitually repress our potential for dialogue. We need look no further than anthropocentric academia for a telling example, as scholars have studied communication for decades as if only humans had voices.

This essay, then, explores a new way of thinking about voice, listening, and nature, integrating and extending scholarly discussions, philosophical theories, and nature writings on this crucial topic. The voice scholar Eric King Watts (2001), for example, argues that there is a proliferation of definitions of voice in rhetorical studies, including the capacity of the speaking subject given the constraints of culture—like the dominant narratives we are born into and within which we are expected to find our place—and the constraints of the linguistic, of language itself. In other words, we don't have complete agency, but we are not completely voiceless. For Watts, voice breaks through when it has an emotional and ethical resonance, and when it is heard. But Watts does not extend his analysis to the emotional and ethical resonance of nature, or more importantly, to Spirit. I will do so, arguing that human voices are more fully liberated, expressing a dialogic wisdom, when the voices of nonhuman nature are more fully liberated and heard.

Thoreau's dialogic wisdom reflects, but pre-dates, Buber's I–thou ontology, in which others may be perceived as sacred Thous rather than monologic Its, including nonhumans. However, Buber's (1957; 1970) ontology of relationship remains dualistic, in the sense that unity experiences are not acknowledged as possible or valuable, while Thoreau, as a transcendentalist, explores the momentary transcendence of dualism in nondual experience. Yet both Thoreau and Buber perceive diverse species as articulating subjects from which we may learn.

Long before Thoreau, of course, pre-neolithic hunter-gatherers embodied this dialogic *attitude*, but such participatory consciousness reflected an animistic orientation, in which everything that is, is alive, rather than transcendent Spirit within all things. This is a significant difference: animism reflects a pre-rational attunement primarily informed by bodily empathy and the senses, and the dialogic wisdom expressed by Thoreau is informed by transrational spiritual experience that transcends and *includes* pre-rational insights, as well as rational knowing. This is not to suggest that hunter-gatherers did not express early forms of thinking in relation to the predator–prey relationship, perceiving places and species as message-bearers from which they received guidance via close observation and mimicry, which was then celebrated and passed on in story and ritual (Shepard, 1997). But unless pre-rational insights are transcended by a more encompassing wisdom, our hunter-gatherer past will be seen as a romanticized golden age to which we must return rather than a stage in human development from which we must learn. This "golden age" included tribalism, short life spans, infanticide as a form of birth control, and propitiation toward unseen forces expressed in ritual sacrifice (Wilber, 1995).

Numerous traditions and thinkers have resisted the reductionism and destructiveness of an I–it, monologic orientation. The Bodhisattva vow of Buddhism, as interpreted by Thich Nhat Hanh, is lived by listening and hearing the "cries of the world in order to come help" (1988, p. 7). The world referenced here goes beyond the human world, since, for Hanh, inter-being is a core principle of an engaged Buddhism, reflecting the spiritual realization that emptiness is form and form is emptiness, and thus reality is ultimately nondual.

Gregory Bateson (1972) did not affirm nondualism, but his work goes a long way toward overcoming our dualistic, and monologic, habits of mind. According to Bateson, our ecological crisis reflects a crisis of perception in which we are living a mistaken epistemology. We are socialized to recognize isolated parts of our world rather than interrelationships among them, sadly unaware that we are immersed in an evolving system of pattern, beauty, and intelligence, which he calls the pattern that connects. Mind and nature are separated rather than integrated, leading to destruction of the communicative system that provides information and sustains our lives.

The ecophilosophies of deep ecology and ecofeminism, and the historical project of ecopsychology, also display a dialogic attitude, extending subjectivity to nonhuman nature. In the Council of All Beings ritual, inspired by deep ecology's championing of a more expansive, ecological self, participants seek insights by speaking from the perspective of diverse species, imaginatively and empathetically voicing concerns over environmental degradation (Seed, Macy, Flemming, & Naess, 1988). In *Woman and Nature* (1978), an early ecofeminist text, Susan Griffin critiques androcentric, transcendentalist philosophies, historically rooted in Platonic idealism, which divorce men from both nature and women, resulting in monologue rather than the ability to listen to nature, hearing "voices from under the earth," wind blowing and trees whispering, the dead singing, and the cries of infants (p. 1). And ecopsychology, in the hands of David Abram (1996), responds to modernist monologue by arguing for an awake and aware perception that recognizes an ongoing pre-verbal reality of silent conversations between our sensing bodies and an animate, more-than-human world. Despite differences—generally speaking, deep ecology's critical focus is on anthropocentrism, ecofeminism on androcentrism, and ecopsychology on the loss of an animistic orientation—all counter I–it attitudes that objectify others, leaving them voiceless and unheard.

Wendell Berry (1990) and Janine Benyus (1997) provide practical expressions of a dialogic attitude. Berry, the Kentucky farmer and progressive Christian, argues for an agricultural ethic in which farmers consult the "genius of the place" in the spirit of dialogue, asking what "nature would be doing if there was no one farming there," what "nature would permit them to do there," and what "they could do there with least harm to the place and to their natural and human neighbors." For Berry, this dialogue is ongoing, leading to "no final accomplishment" that can be foreseen (p. 209). In other words, we must keep the conversation going, remaining open to new insights. Benyus, in *Biomimicry*, argues for a deep listening in which nature is seen as model, measure, and mentor, informing our technological practices, including in agriculture, the harnessing of energy, and healing. Nature, through generations of evolution that has generated what works—failures are fossils—has provided design solutions to many of our problems; that is, if we are able to listen, opening ourselves to the patterns that connect and receiving information.

Increasingly, academics have been arguing for the necessity of practicing a dialogic attitude characterized by nonverbal listening in response to ecocrisis. In rhetorical studies, Maxcy (1994) argues for a phenomenological approach that uncovers our entanglement in nature and nonlinguistic meanings. Nature may speak when we bracket the "natural attitude" (as in doxa), suspending dominant discourse, beliefs, and habits of perception by directing our "attention to how 'things' present themselves within the stream of our intentional life" (p. 336). Such bracketing and attentiveness leads us to a heightened awareness of the perceptual process, which is neither objective nor subjective but relational, allowing us to experience nature anew, combining insights from symbolic action with previously unrealized experience.

Rogers (1998) criticizes constitutive approaches to knowing nature, as this approach maintains the mind–nature dualism, arguing for the possibilities for transhuman dialogue in which we come to know ourselves via interactions with diverse nature that call for an ethical response. For Rogers, social construction theories are inevitably monologic, since they assume that nature cannot be experienced trans-lingua, beyond words and thoughts. Nature is a diverse, material, affective force existing outside of language that may provide instruction, while not being fully known. Rogers' materialistic, transhuman, dialogic theory of communication argues that we must learn to listen to "nondominant voices and nonhuman agents," but he does not provide any specific practices of how to do so (p. 268). He also rejects the spiritual dimension of life, fearing that it is too totalistic.

Salvador and Clarke (2011) argue for an embodied critical rhetoric informed by the weyekin principle, based on the Nez Perce experience of being empowered by the strength of a specific animal via a vision quest, and from which we may continue to learn by listening via resonance, or "bodily meaning beyond the symbolic," observation, and mimicry (p. 249). For Salvador and Clarke, the weyekin principle provides the practices missing from Rogers' theory of transhuman dialogue. These practices are materialistic and corporeal rather than metaphysical or spiritual, and thus there is no transcending of nature through symbolization, which they argue is a key problem, as language and constitutive theories block our embodied, nonverbal experience of nature and nonhumans. The suggestion here is that we must experience and then symbolize, rather than vice versa, as dominant systems of meaning limit much of our potential experience (p. 248). This is true, but the spiritual dimension of life is once again poorly articulated or denied as irrelevant to listening to nature's voices.

In environmental ethics, Friskics (2001) argues for a more encompassing view of logos, suggestive of the ancient Greek view which bound the rational soul to the cosmos, and thus of rational and contemplative inquiry. Logos bespeaks "the way of things," and the Word made flesh as primal speech, song, and voice "in and through the beings and things of creation" (p. 393). Everything speaks via their presence, and thus everything participates in logos. Friskics defends his position via Buber, but his views are ultimately rooted in his experience. He begins his essay with a description of his rustic cabin home near the east slope of the Montana Rockies, and then proceeds to describe his daily walks to Crown Mountain through forest, hillside, and grassy knoll, which he does for typical reasons—to stretch his legs, get some sun, check the weather—but also because he feels called by the mountain. Crown Mountain "speaks" to Friskics via its "silent, hulking presence," and he feels compelled to respond (p. 392).

Much of this needed and timely discussion—and yet late from the perspective of various ecocrises—is occurring among environmental communication scholars (Carbaugh, 1999; 2007; Milstein, 2008; Salvador & Clarke, 2011). Carbaugh (1999), for example, studied the Blackfeet practice of "just listening," in which tribal members willfully, and then humbly, become co-present and co-participant with the spirits that inhabit places and species, conversing with ancestors, keeping traditions alive, expanding self and awareness, and learning what they need to learn.

Environmental communication scholarship is critical of the academic habit of focusing on rhetoric at the expense of the actual world, or the rational focus on the social construction of nature via discourse rather than the continual construction of an embodied and fluid and listening self attuned to the natural world that then discourses. Criticizing the power of dominant discourses—like the discourses of bigger, better, and more—to shape monologic, I–it perceptions of nature is obviously needed, but more steps are necessary. Or rather, as Carbaugh (2007) puts it, we must focus on "representations of nature" but also on its "presentations," becoming attuned to other "expressive systems" (p. 68).

These are just some of those advocating for dialogic relations with nature (see also Deloria, 1999; Fisher, 2002; Greenway, 1995; Hogan, 1995; Kaza, 1993; Keller, 1983; Martin, 1992; Rezendes, 1999; Roszak, 1992; and Sewall, 1999). All make a contribution; however, most of the scholars, philosophies, and nature writers mentioned argue for a pre-rational return to a dialogic attitude attentive to the nonverbal via the body and senses, or champion rational inquiry that questions dominant narratives and perceptual habits that exclude the nonverbal, or nonlinguistic modes of knowing. Some provide useful practices for listening to nature's voices (biomimcry, bracketing, self-inquiry and transformation, etc.) but all, except Buddhism, leave out the transrational. Again, this is not a small difference. The championing of pre-rational experience may lead to anti-rationality and questionable solutions to ecocrisis, like the desire to deindustrialize to a point prior to the birth of agriculture. On the other hand, championing rationality, or at least instrumental rationality, typically leads to techno-industrial solutions that lead to more problems, like the countless abuses of factory farming. What we need is worldcentric rational dialogue on the crisis of agriculture, global warming, and our many inter-related eco-social crises, informed by insights and the call to responsibility incurred from pre-rational, rational, and transrational experience, not a mere return to pre-rational modes of knowing or the valorizing of rationality. Denigrating rationality and ignoring transrational experience, or focusing on rational inquiry and ignoring pre-rational and transrational experience, or confusing the transrational with the pre-rational, also leaves us without a genuine path of spiritual development, in which humans grow, in fits and starts, through stages of pre-rational, rational, and transrational awareness (Wilber, 1995).

Thus, while there has been much needed discussion of dialogic relations beyond the human, and some action inspired by this discussion (sometimes ill-informed), there is a reason why Thoreau is a seminal figure of environmental concern: he embodied, in his life path and daily practice, not only a dialogic attitude but dialogic wisdom informed by the integration of multiple practices and modes of knowing. Nature speaks, or is silenced, in relation to what we bring to our encounters and interactions: by our ability to bracket dominant narratives and habits of perception, by our openness to multiple modes of knowing nature and nonhuman others, exploring expressive systems that include the nondual, and, ultimately, by the range and quality of our verbal and nonverbal competence.

Thoreau clears a path for counter-narratives that integrate mind and nature. He does so in his writings, redefining human development within a spiritual and earth context and our capacity for attending to both diverse voices and the voice of nature, but also by living them, by living part and parcel of nature. We must listen, then, in the manner of Thoreau, to our senses, to rational inquiry and our poetic imaginations, and to spiritual experiences. We must listen, as he writes in the "Sounds" section of *Walden*, to "the language that all things and events speak without metaphor" (1982, p. 363).

Cultivating nondual awareness: The voices and voice of nature

In the "Higher Laws" section of *Walden*, Thoreau makes the following claim: "Nature is hard to overcome, but she must be overcome" (1982, p. 462). This is a startling statement, especially since the majority of his writings record in considerable detail his sensuous and spiritual connection with nature. In this statement, however, Thoreau is expressing his sympathy with transcendentalist principles that seem to privilege Spirit over matter. Emerson (in Emerson & Thoreau, 1991), for example, wrote of nature: "Through all its kingdoms, to the suburbs and outskirts of things, it is faithful to the cause whence it had its origin. It always speaks of Spirit" (p. 41). But while Thoreau's vision encompasses the transcendent, his focus is consistently on the immanent: "My profession is to be always on the alert to find God in nature, to know his lurking places, to attend all the oratorios, the operas, in nature" (1961, p. 58). This is a telling passage for someone who spent much of his youth searching for vocation, which he eventually found as a writer who listened to the songs of nature, hearing both Spirit and diverse voices, while criticizing the deaf ears and desperation of the mass of men. References to nature's music occur often in his writings, whether the "ambrosial" creaking of crickets, the "cool bars of melody" of the wood thrush, or the mystical tinkling of his hoe against a rock while tending his bean field (1961, pp. 91–92). For Thoreau, the voice (Spirit) within nature and voices of nature are a balm to human awareness, awakening our potential for transcendentalist "self culture," or the call to spiritual development.

Thoreau's commitment to spiritual development is made explicit in numerous passages of *Walden*. In the opening section, "Economy," he argues that true philosophers are "progenitors of a nobler race of men" (1982, p. 270), and in a later passage he turns to nature for a symbol of the human condition, comparing the majority of humankind to a torpid snake:

> I saw a striped snake run into the water, and he lay at the bottom, apparently without inconvenience, as long as I stayed there, or more than a quarter of an hour; perhaps because he had not yet fairly come out of a torpid state. It appeared to me that for a like reason men remain in their present low and primitive condition; but if they should feel the influence of the spring of springs arousing them, they would of necessity rise to a higher and more ethereal life.
>
> (1982, p. 296)

Thoreau's devotion to the "spring of springs," or our developmental potential, is also declared in the aptly named "Where I Lived, and What I Lived For," in which he compares the morning to spiritual awakening: "Morning is when I am awake and there is a dawn in me ... The millions are awake enough for physical labor; but only one in a million is awake enough for effective intellectual exertion, only one in a hundred millions to a poetic or divine life" (1982, pp. 342–343). Thoreau goes on to state that the poetic or divine life requires effort:

> I know of no more encouraging fact than the unquestionable ability to elevate his life by a conscious endeavor. It is something to be able

to paint a particular picture, or to carve a statue, and so to make a few objects beautiful; but it is far more glorious to carve and paint the very atmosphere and medium through which we look, which we morally can do. To affect the quality of the day, that is the highest of arts.

(1982, p. 343)

Thoreau's focus on spiritual awakening, and a heightened clarity of perception, is forcefully stated, although he softens his pessimism concerning the "millions," allowing for the possibility for development when it is made a conscious endeavor. For Thoreau, listening to "the oratorios, the operas" in nature is a first conscious step toward an elevated awareness, and this requires receptive senses. We affect the quality of the day through the disciplined and incessant use of seeing, hearing, touching, tasting, and smelling nature's bounty in the present moment. This practice then paves the way to contemplation of and communing with the eternal, with higher laws expressed as spiritual unity amid the diversity revealed by the senses: "In eternity there is indeed something true and sublime. But all these times and places and occasions are now and here. God himself culminates in the present moment, and will never be more divine in the lapse of ages" (1982, p. 349).

Thoreau experiences this eternal now most revealingly in the hoeing of his bean fields. He is often critical of meaningless work, including the business of agriculture, as it diverts too much time away from pursuits like walking and contemplation, but he has affection for his bean field that he discusses as a dialogic relationship: "What shall I learn of beans or beans of me? I cherish them, I hoe them, early and late I have an eye for them; and this is my day's work" (1982, p. 405). Thoreau soon discovers that hoeing and contemplation are compatible activities, and this becomes a stepping stone to nondual experience: "When my hoe tinkled against the stones, that music echoed to the woods and the sky, and was an accompaniment to my labor which yielded an instant and immeasurable crop. It was no longer beans that I hoed, nor I that hoed the beans..." (1982, p. 408).

In this passage, Thoreau experiences a momentary loss of separate self. While there is physical distance, and thus relationship, there is no psychological division between Thoreau and the beans, or Thoreau and his work, and in this reverie his actions express an effortless effort. Thoreau is the best "bean-hoer" on the block. There is no past, no future, only the eternal now of right action. Throughout *Walden* Thoreau repeatedly calls for us to simplify our lives so that we don't waste time on inessential activities. In this example, however, his call to "simplify, simplify, simplify" takes on its ultimate spiritual meaning—we are called to the simple, clear awareness revealed in nondual experience. And, as any Zen master will claim, there is nothing extra-ordinary about this awareness. Rather, it is our ordinary awareness when our perceptions are not filled with the habitual clutter of unproductive desires. Desire, like the desire to affect the quality of the day, is not our enemy; in fact, it is the precondition of much communicative praxis, whether with humans or nonhumans. We crave, and need, connection with others. But there are moments of transcendence, or the overcoming of desire, in which our responsiveness reaches new heights.

Nondual experience leads to the transcendentalist realization that we are Spirit aware of Itself. This experience of unity-in-diversity transcends and includes insights from attending to our senses and listening to the sounds, and silences, of nature, as well as the poetic and aesthetic contemplation of unity-in-diversity. It could be argued that such experience is translogical rather than dialogical—we have transcended the logic, and illogic, of dualism—but in nondual moments a material communicative relationship remains. Nondual experience does not negate others, it recognizes that others are also us—thou art that. Such experiences are communicative because an insight, the unity of all things, has been received with clarity, and the profundity and power of this insight cannot be discovered in any other way. And it only takes a glimpse for our lives to be informed and transformed. Thus, if we read "overcoming nature" within the context of Thoreau's glimpses into nondual awareness, as well as his claim that he "loves the wild not less than the good," it is not really nature that he wishes to overcome, but an understanding of nature that does not also acknowledge Spirit (1982, p. 457). Listening to the diverse voices of nature, and their diverse lessons, has led to a realization: everything speaks of Spirit.

For Buber, this argument for unity means that "there is no longer anything over and against him," and the "great dialogue between I and Thou is silent" (1957, p. xvi). But silence speaks volumes, providing an insight that is part of human experience. Buber, as a younger man, affirmed the profundity and power of unity in experience, but then moved from mysticism to dialogue. Thoreau, who called himself a mystic, turned to science in later life, perceiving unity-in-diversity. Interestingly, both Buber and Thoreau extol the heightening of our experience of this world.

A sympathy with intelligence

Thoreau is a grounded transcendentalist, sharing his spiritual practice: "simplify, simplify, simplify," both mentally and materially, living deliberately; read and write; saunter amid nature, attending to the senses and voices of nature; and aesthetic-poetic contemplation of Spirit, of the voice of nature perpetually shining through diverse nature. Such practice may lead to transrational experience, transcending and including insights from attending to the senses and contemplation, but in "Higher Laws" he suggests such knowing has limits: "Nature is a personality so vast and universal that we have never seen one of her features" (1982, p. 625). And so, Thoreau confounds with a contradiction: we must be spiritual seers and yet nature's personality is beyond being seen. Or, in the context of this essay, we must practice dialogic wisdom but there are limits to dialogue and wisdom, to our ability to read the language which all things speak. He responds to this contradiction, in his essay "Walking," when he writes: "The highest we can attain is not Knowledge, but Sympathy with Intelligence" (in Emerson & Thoreau, 1991, p. 113). Thoreau chooses a wise path: not complete knowing or not-knowing, but sympathy with intelligence expressed as the receiving of insights from multiple modes of knowing.

There are many passages in *Walden* that illustrate sympathy with intelligence, but the opening paragraph of "Solitude" is one of the most poetic:

> This is a delicious evening, when the whole body is one sense, and imbibes delight through every pore. I go and come with a strange liberty in Nature, a part of herself. As I walk along the stony shore of the pond in my shirt sleeves, though it is cool as well as cloudy and windy, and I see nothing special to attract me, all the elements are unusually congenial to me. The bullfrogs trump to usher in the night, and the note of the whippoorwill is borne on the rippling wind from over the water. Sympathy with the fluttering alder and poplar leaves almost takes away my breath; yet, like the lake, my serenity is rippled but not ruffled.
>
> (1982, p. 380)

This is a beautiful articulation of the sensuous and spiritual. It is a "delicious evening, when the whole body is of one sense," which echoes synaethesia, or the overlapping of senses leading to a heightened aesthetic awareness. Thoreau experiences the divine in sympathy with the trees, with the body, with the senses: all are integrated, revealing a "strange liberty within Nature" in which a spiritual unity amid material diversity is initially apprehended.

Thoreau's sympathy with intelligence also included the practice of science, to which he turned in later life. In *No Man's Garden: Thoreau and a New Vision of Civilization and Nature*, Daniel B. Botkin (2001) uses the example of Thoreau measuring the depths of Walden Pond to show his early and adept scientific practice. Using a compass, sounding line, and weight, Thoreau made over 100 measurements which he translated into a map. Botkin writes that his curiosity concerning his quantitative measurements led him to formulate hypotheses about other ponds that required further measurements and hypotheses (pp. 65–67). Thoreau followed the scientific method, and Botkin argues that facts brought him closer to nature, as well as causing him to praise the ways of science: "Science is always brave; for to know is to know good; doubt and danger quail before her eye" (p. 75). However, Thoreau's scientific practice also included collecting species for Louis Agassiz, a leading scientist at Harvard, and the "bravery" of science was not always apparent, as he records his killing of a box turtle in his journal: "I cannot excuse myself for this murder, and see that such actions are inconsistent with the poetic expression... I pray I may walk more innocently and serenely through nature" (1961, p. 135). In another section of his journal, he puts the matter bluntly: "The man of science, who is not seeking for expression but for a fact to be expressed merely, studies nature as dead language" (1961, p. 112).

Thoreau's scientific practice was mostly poetic rather than objective. He acknowledged the unavoidable subjective dimension of science, arguing that the biography of the scientist reveals more than what is studied. This was both a rational criticism and a deeper truth, or a more complete insight into reality and living nature. As a result he was quite fond of Goethe, both for his literature and his investigations into plant metamorphosis. Goethe practiced a subjective involvement with the plants under study, and it led him to recognize the law of the leaf, or the directing and animating life force underlying matter. Goethe was interested in apprehending the growing process and pattern and intelligence that manifested itself throughout the plant. Thus he perceived that the pattern of the leaf was present in the plant as a whole, and by extension, in nature as a whole (Richardson, 1986).

In *Natural Life: Thoreau's Worldly Transcendentalism*, David M. Robinson (2004) writes that Thoreau's later scientific practice must be interpreted within the context of Goethe's law of the leaf, as well as transcendentalism and the poetic imagination, to which he still adhered, even though his focus was increasingly on cataloguing nature's diversity. He admired the "old naturalists" who "sympathized with the creatures they described" (p. 182). According to Robinson, Thoreau's naturalist tendencies led him to recognize "effluence," a concept he derived from attending to the "fragrance of fruits and blossoms." Effluence linked the observer with the observed, making them receptive to the expressions or utterances of the species under study. Thoreau wrote in his journal: "Only that intellect makes any progress toward conceiving of the essence which at the same time perceives the effluence" (as cited in Robinson, 2004, p. 181).

Thoreau's scientific practice was rigorous. In fact, his dated records of wildflower blossomings are being used by current scientists to track climate change (Levy, 2012). Yet his practice was also marked by sympathy, and, thus, the language that all things speak includes information received from scientific investigation. Thoreau's spiritual practice also includes, then, the rational yet empathetic orientation of the naturalist, but within the context of nondual experience. It is not surprising that he wrote: "Perhaps the facts most astounding and most real are never communicated by man to man" (1982, p. 463).

We need not have nondual experiences to listen to the voices of nature, receiving telling information on the primacy of relationship— receptive senses and contemplation, as well as scientific practice, may overcome the rigid dualism between mind and nature—but nondual glimpses more fully place us in sympathy with intelligence, with the voice of Spirit within and expressed by nature, guiding our life path by making us receptive to multiple modes of knowing. Thus, Thoreau's life and legacy are ultimately defined by his spiritual vision, such as in this passage from "The Ponds" in which he poetically describes via an analogy with fishing the nondual integration of transcendent and immanent dimensions:

> It was very queer, especially in dark nights, when your thoughts had wandered to vast and cosmogonal themes in other spheres, to feel this faint jerk, which came to interrupt your dreams and link you to Nature again. It seemed as if I might cast my line upward in the air, as well as downward into this element, which was scarcely more dense. Thus I caught two fishes as it were with one hook.
>
> (1982, p. 424)

Speaking for nature?

Thoreau's spiritual vision led him to consistently practice dialogic wisdom, but the first line of "Walking," while seeming to be a straightforward celebration of freedom, wildness, and identification with nature, raises critical questions: "I wish to speak a word for Nature, for absolute freedom and wildness, as contrasted with a freedom and culture merely civil—to regard man as an inhabitant, or part and parcel of Nature, rather than a member of society" (Emerson, 1991, p. 71). The simple phrase, "I wish to speak a word for Nature," is also a combustible one. Can Thoreau speak *for* nature? Should he even try, or is the attempt an act of hubris reflecting monologic domination? Or perhaps it is simply destined for failure? We may be part and parcel of nature, but when we speak and write we speak and write as humans, not nature—right?

In "Speaking for Nature: Thoreau and the Problem of Nature Writing," Nancy Craig Simmons (2000) argues that claiming to speak for nature imposes ourselves on nature, controlling and reducing it via language for human consumption and use. This is hardly the way we typically think of Thoreau, but Simmons raises difficult questions. Can we really perceive nature directly via spiritual experience, or the senses, or is it always socially constructed, with humans inevitably describing, translating, recording, reporting, and interpreting? In other words, is language ever transparent, revealing the world of non-words with the world of words, or is it always laden with cultural meanings that keep us separate and apart rather than part and parcel of the natural world?

I have turned to Thoreau as an exemplar of dialogic wisdom informed by nondual experience, and yet Simmons' criticisms deserve attention. The shadow side of nondualism, or claims for nondual experience, are claims of literally channeling nature's voices. In other words, if nature writing can be transparent, wouldn't this group Thoreau with New Agers who claim they can commune with crystals and channel dolphins and dead yogis, providing literal knowledge rather than sympathy with intelligence? Such literalness must be resisted, as should the knee-jerk counter-claim that language and communication are human constructs placing an unbridgeable gulf between nature and humanity. After all, nondual experiences have been documented in all cultures, language also emerged from the earth, or as Emerson (in Emerson & Thoreau, 1991) declared, "Words are signs of natural facts," and our bodies and psyches participate in ongoing nonverbal communication with the nonhuman world (p. 22).

For Thoreau, transcendentalism presumes a meaningful natural world saturated with Spirit, of which we are part and parcel, and thus we can sympathetically interpret the unfolding of new and vital meanings, letting nature speak in experiences of reciprocity, not in literal translations. Thus, sympathy with intelligence allows us, in our better moments, to "speak a word for nature" by transcribing our discoveries. Simmons, on the other hand, finds such sympathy to be a manifestation of instrumental use, or another monologue. Encountering nature spiritually still leads us to use nature for our benefit. In fact, she states that Thoreau habitually uses nature in his writing for poetic purposes. Thus he does not let nature speak; rather, he makes use of nature for the sake of symbol, language, and the craft of nature writing, creating a dualism between mind and nature.

Simmons (2000) does praise some of Thoreau's writings, particularly his journal description of tracking a fox for several miles. A light snow has betrayed the path of the fox and Thoreau follows and records its movements. The writing is pure observation and detail, but then ends with the following: "How dangerous to the foxes and all wild animals is a light snow...betraying their course to hunters...I followed on this trail so long that my thought grew foxy" (as cited in Simmons, 2000, p. 231). Simmons praises this entry because to think like a fox, and to let it speak through writing (to the degree that one can), demands close observation and recording, not poetic flourishes or inspired spiritual heights that suggest anthropomorphism.

Simmons (2000) further argues that Thoreau himself was concerned with the difficulty of speaking for nature. In *Walden*, he writes:

> I fear chiefly lest my expression may not be extra-vagant enough, may not wander far enough beyond the narrow limits of my daily experience, so to be adequate to the truth of which I have been convinced...I desire to speak somewhere without bounds; like a man in a waking moment, to men in their waking moments.
> (as cited in Simmons, 2000, p. 230)

Simmons states that this passage reflects Thoreau's desire to speak beyond the limits and boundaries of culture. And for her, he comes closest when recording events like the path of the fox. Field notes, then, tend to have a greater ability to let nature speak, as they supply a running record of our observations, and are not writing crafted for human use. But this interpretation assumes that there is nothing to be gained from reflecting on experience and carefully crafting our language. And a "running record" does not remove cultural habits of thinking and seeing; rather, it unthinkingly reproduces an objective, scientific bias.

Thoreau is clearly referring to spiritual awakening by wishing to speak extra-vagantly. He desires to share insights from nondual experiences via his writing to others who are also spiritually awake and aware, or to stimulate others to be awake and aware. But Simmons dismisses Thoreau's spiritual writings as reflecting the bias of transcendentalism, as they reflect Emerson's claim that nature is a symbol of Spirit, and a symbol can only be used as a writer's tool. However, this response, while not without critical insight concerning the dangers of transcendence without immanence (i.e., nature is only a symbol of Spirit) fails to account for the evocative power of Thoreau's spiritual writings, the possibility that both Spirit and nature are given voice, to some degree, via poetic prose, and that while symbols may divide us from nature, they need not always lead to dualism. Or, as Carbaugh (2007) argues, the environment is doubly quoted, and both "the word and the world speak" (p. 64).

Thoreau's commitment to and experience of the language that all things speak is expressed often in his writings: we are called to the simple, clear awareness of awakened consciousness and to listen to nature's chorus. The passages cited in this paper only scratch the surface of Thoreau's deep inquiry into the relations between Spirit and nature, between the voice of unity and diverse voices, yet these passages speak with power and immediacy because of Thoreau's craft as a writer. He attempts, as mystics always have, to communicate "extra-vagantly," articulating nondual experiences that are beyond words and thoughts with words and thoughts—an attempt that inevitably falls short, but that may inspire others to spiritual practices, to listening to the voices and voice of nature.

Simmons, on the other hand, criticizes the "transcendentalizing" of nature, concerned that the particular (voices) will be lost in the totalizing of experience (Voice). But she clearly does not acknowledge the possibility of nondual experience, and that the universal may be found in the particular, or that the particular may lead to the universal, and to an even deeper reverence. Still, Simmons' core questions remain: can we speak for nature? What constitutes good nature writing? And is it possible to not use nature when writing? It is more important to ask these questions than respond with rigid answers. Still, I would respond by agreeing that we are cultural animals, and thus writing is never fully extra-vagant, or beyond the bounds of language. But this merely means that nature writing should interpret via multiple forms of language, including scientific-naturalistic field notes, story, critical rationality, contemplative/poetic-aesthetic, and spiritual, and thus multiple forms of experience and perception. Sympathy with intelligence demands nothing less. If such sympathy constitutes use, then there are clearly different types of use, with the listening and writing practiced via embodied virtues like humility, empathy, care, and reverence reflecting a more responsive use. We merely need to begin a practice of listening, comparing our findings with those of other competent listeners.

Thoreau attended to nature like few others, developing skills of observation, embodied virtues, and spiritual reflection. If anyone can speak a word for nature, blurring the boundaries between science and poetic imagination, he can. Language is always of the world, and thus there is always a cultural bias, but Thoreau's bias is integrative, informed by experience beyond words and thoughts while remaining open to not-knowing and free from totalization, and thus it reflects a better bias. And anthropomorphism, when abused rather than reflecting an empathy that reveals relational insights, should be questioned via critical rationality informed by other modes of knowing. Sympathy with intelligence does not lead to anthropomorphism, but to an awake and widened awareness.

Conclusion: The call to responsibility

Thoreau was a complex man attempting to express a diversity of experience. He occasionally isolated himself from townspeople, and was capable of emotional detachment when observing nature, but there is little doubt that he experienced the depths of solitude, discovering the further reaches of his spiritual nature. However, despite having friends, his interpersonal intelligence may have needed development. Thoreau often uses blunt prose when speaking of people and society and poetic prose when speaking of, or for, nature. Emerson stated in his eulogy for Thoreau: "There was somewhat military in his nature not to be subdued, always manly and able, but rarely tender, as if he did not feel himself except in opposition" (1982, p. 7). On the other hand, it is a falsehood that Thoreau was a hermit; in fact, it is more accurate to describe him as a "townie" intimately involved in the social life of Concord, whether teaching or playing with children, walking in the woods with friends, engineering new machinery and pencil designs for his father's pencil business, surveying, giving speeches, or conversing with neighbors and fellow transcendentalists on key causes of the day, including the abolitionist movement (Sullivan, 2009). Still, Thoreau's identification with nature seems to be stronger than his connection with people, but often for understandable reasons. He saw the destructive potential of industrialization more clearly than his fellows, who busied themselves with the business of making a living rather than a life of continuous awakening. And yet, it is these very complexities of his life in Concord, and his personality, that led him to the Walden experiment, to write, walk, and contemplate, and to discover a solace and spirit in the natural world that few have equaled.

Thoreau's main influence is his quest for spiritual development. This quest, which was embraced by listening to Spirit and nature, to the language that all things speak, as well as to his inner nature or genius, is what makes him an influential figure. Thoreau's life is complex and resoundingly unique, but his pursuit of development and many struggles have universal appeal. Like all of us, Thoreau failed at continuously living this spiritual awareness. But rather than this making him a failure, it merely makes him human. Thoreau, the poetic champion of nature, also accidentally set fire to the woods (Sullivan, 2009).

Thoreau still speaks to us today because of the universal experiences he explored, but also because our narratives of growth, technology, and progress, of bigger, better, and more, need to be redefined within the context of attentive listening to the voices and voice of nature, integrating insights from pre-rational, rational, and transrational experience, transcending the dualism between mind and nature. Such practice raises a critical question, however: do toxic waste and polluted rivers also speak of Spirit? Are the voices of nature still present? Yes and no. Spirit is within every material manifestation, but humans obviously have the destructive ability to ignore the ethical constraints suggested by this realization, extirpating or chasing away the voices of diverse species. Yet, recognizing Spirit leads us to more clearly see what was, and the potential of what could be. Thus, listening to the voice and voices of nature has a practical result: we will refrain from doing damage in the first place, and we may more wisely restore damaged habitat by re-storying the land, attending, as best we can, to the voices that once were. Still, as Awiaka, a Cherokee poet, writer, and activist, articulated, it is "difficult to listen through concrete" (Carbaugh, 2005, p. 118).

Most certainly there are constraints on voice, and yet there is much that can be done to liberate nature's voices as well as our own, allowing them to be heard. Such liberation provides us with an expanded array of responses, whether confronting polluted rivers or discussing global warming, as differing modes of knowing and insights are called forth in particular situations, informing our arguments and unfolding life choices. Carbaugh (2007), for example, argues that discursive diversity is essential, as is our ability to weave the best discourses together, "within places well known" (p. 70). No single perspective is final, and there will always be fruitful contradictions that resist situational integration, but this merely means that we must keep the conversation going, including with nonhuman others. Rather than being voiceless, then, we may embody a widened awareness, recognize that living means being addressed, and expand our capacity for listening and agency, which includes a critical assessment of our ecological predicament. Each mode of knowing dialogically informs the other, providing support for worldcentric dialogue that assesses the excesses of industrialization and empowers us to speak out against the silencing of nature's voices.

In *Walden*, Thoreau writes: "We do not ride on the railroad; it rides upon us" (1982, p. 345). And in his journal he laments the extirpation of species: "Is it not a maimed and imperfect nature that I am conversant with?... I listen to a concert in which so many parts are wanting" (1961, p. 157). These sentiments are not of the past; they deeply reflect our current predicament. Thoreau offers the wood thrush's song in response, which tells the story of the "immortal wealth and vigor" of the forest; when we hear it, the "gates of heaven are not shut" such that new worlds may be created and institutions amended (1961, pp. 92–93).

May we learn to listen to the voices and voice of nature, rethinking our monologic myths. May we hear the call to responsibility and action.

References

Abram, D. (1996). *The Spell of the Sensuous: Perception and Language in a More-Than Human World*. New York: Pantheon.
Bateson, G. (1972). *Steps to an Ecology of Mind*. New York: Ballantine.
Benyus, J. (1997). *Biomimicry: Innovation Inspired by Nature*. New York: Quill.
Berry, T. (1988). *The Dream of the Earth*. San Francisco: Sierra Club Books.
Berry, W. (1990). *What Are People for?* San Francisco: North Point Press.
Botkin, D. B. (2001). *No Man's Garden: Thoreau and the New Vision for Civilization and Nature*. Washington, DC: Island Press.
Buber, M. (1957). *Pointing the Way*. New York: Harper & Row.
Buber, M. (1970). *I and Thou*. Ed. Walter Kaufmann. New York: Scribner's.
Carbaugh, D. (1999). "Just Listen": "Listening" and landscape among the Blackfeet. *Western Journal of Communication, 63(3)*, 250–270.
Carbaugh, D. (2005). *Cultures in Conversation*. New York: Routledge.
Carbaugh, D. (2007). Quoting "the environment": Touchstones on Earth. *Environmental Communication: A Journal of Nature and Culture, 1(1)*, 64–73.

Deloria, V. Jr. (1999). *Spirit and Reason*. Colorado: Fulcrum Pub.
Emerson, R. W., & Thoreau, H. D. (1991). *Nature/Walking*. Ed. John Elder. Boston: Beacon Press.
Fisher, Andy. (2002). *Radical Ecopsychology: Psychology in the Service of Life*. New York: SUNY Press.
Friskics, S. (2001). Dialogic relations with nature. *Environmental Ethics, 23(4)*, 391–410.
Greenway, R. (1995). The wilderness effect and ecopsychology. In T. Roszak, M. E. Gomes & A. D. Kanner (Eds.), *Ecopsychology: Restoring the Earth, Healing the Mind* (pp. 122–35). San Francisco: Sierra Club Books.
Griffin, S. (1978). *Woman and Nature: The Roaring inside Her*. New York: Harper & Row.
Hanh, T. N. (1988). *The Heart of Understanding*. Berkeley, CA: Parallax Press.
Hogan, L. (1995). *Dwellings*. New York: Touchstone.
Kaza, S. (1993). *The Attentive Heart: Conversations with Trees*. New York: Ballantine.
Keller, E. F. (1983). *A Feeling for the Organism*. New York: W. H. Freeman.
Levy, R. (2012, March 14). Thoreau's journals help scientists track climate change. Retrieved on December 3, 2013 from http://slatest.slate.com/posts/2012/03/
14Martin, C. L. (1992). *In the Spirit of the Earth*. Baltimore: John Hopkins University Press.
Maxcy, D. J. (1994). Meaning in nature: Rhetoric, phenomenology, and the question of environmental value. *Philosophy and Rhetoric, 24(4)*, 330–346.
Milstein, T. (2008). When whales "speak for themselves": Communication as a mediating force in wildlife tourism. *Environmental Communication: A Journal of Nature and Culture, 2(2)*, 173–192.
Rezendes, P. (1999). *The Wild Within*. New York: Berkeley Books.
Richardson, R. D. Jr. (1986). *Henry Thoreau: A Life of the Mind*. Berkeley, CA: University of California Press.
Robinson, D. M. (2004). *Natural Life: Thoreau's Worldly Transcendentalism*. Ithaca, NY: Cornell University Press.
Rogers, R. A. (1998). Overcoming the objectification of nature in constitutive theories: Toward a transhuman, materialist theory of communication. *Western Journal of Communication, 62(3)*, 244–272.
Roszak, Theodore. (1992). *The Voice of the Earth*. New York: Simon & Schuster.
Salvador, M., & Clarke, T. (2011). The Weyekin principle: Toward an embodied critical rhetoric. *Environmental Communication: Journal of Nature and Culture, 5(3)*, 243–260.
Seed, J., Macy, J., Flemming, P., & Naess, A. (1988). *Thinking like a Mountain: Towards a Council of All Beings*. Philadelphia: New Society Publishers.
Sewall, L. (1999). *Sight and Sensibility: The Ecopsychology of Perception*. New York: Tarcher/Putnam.
Shepard, P. (1997). *Coming Home to the Pleistocene*. Washington, DC: Island Press.
Simmons, N. C. (2000). Speaking for nature: Thoreau and the problem of nature writing. In R. Schneider (Ed.), *Thoreau's Sense of Place: Essays in American Environmental Writing* (pp. 223–234). Iowa City: University of Iowa Press.
Sullivan, R. (2009). *The Thoreau You Don't Know*. New York: HarperCollins.

Thoreau, H. D. (1961). *The Heart of Thoreau's Journals*. Ed. Odell Shepard. New York: Dover Pub.
Thoreau, H. D. (1982). *The Portable Thoreau*. Ed. Carl Bode. New York: Penguin.
Watts, E. K. (2001). "Voice" and "voicelessness" in rhetorical studies. *Quarterly Journal of Speech, 87(2)*, 179–196.
Wilber, K. (1995). *Sex, Ecology, Spirituality: The Spirit of Evolution*. Boston: Shambhala Publications.

10
The Ethics of Listening in the Wilderness Writings of Sigurd F. Olson

David A. Tschida

As environmental advocates work to advance an environmental ethic, many have turned their attention to historic writers and activists (e.g., Thoreau, Muir, Leopold, and Carson) in the hope of recovering some lost understanding, value, or spirit that will add to that effort. This chapter examines one such figure whose contribution has been overlooked but whose significance to this book's theme encourages new attention. Sigurd F. Olson (1899–1982) was the author of nine books and many essays for the nation's leading newspapers and outdoor-themed magazines. His book *Wilderness Days* earned him the 1974 John Burroughs Medal for nature writing. He was an active speaker in his role as the president of the National Parks Association and the Wilderness Society and as an advisor to the National Park Service and Stewart Udall, a U.S. Secretary of the Interior. Of Olson, Backes (2001) contends, "[He] was not the first American to discuss the spiritual value of wilderness, nor was he the most scholarly. He was simply the most beloved advocate of his generation" (p. xix).

In this chapter, I examine how Olson challenges our accepted ideology about a common human-to-human communicative practice—dialogic listening—by identifying it also in the context of nature-to-human communication. My objective is to extend the ethics of human-to-human dialogic listening to include the listening we perform with nature as an agent or Other—as dialogic partner. To reach this end, I explore one aspect of the traditional taxonomy of listening purposes and processes (e.g., Wolvin & Coakley, 1993; Purdy, 1991); a taxonomy with an influence so powerful that I argue it has functioned within communication scholarship and everyday practices of listening as an ideology privileging human-to-human communication to the neglect (if not rejection) of other listening applications. Wolvin and Coakley (1993) describe a human-centered typology typical of this ideology that they divide into five generalized listening purposes. *Discriminative* listening is the work done to distinguish one sound from another. *Comprehension* is the labor involved in understanding a speaker's message. *Therapeutic* is giving attention to speakers who are trying to make sense of their own thoughts or feelings by talking them through with a listener. *Critical* listening describes the energy given to form a judgment about a speaker's message. Finally, *appreciative* listening is the enjoyment or pleasure that comes to an audience from music or, as Wolvin and Coakley note, nature.

A typology like this arises from and reproduces an ideology about communication practices that ignore or reject the possibility that nature can be an agent or subject/Other worthy of our listening, positioning nature instead of an object that exists to pleasure our senses. Since dialogic listening falls under the comprehension-centered purpose or practices, extending its scope to recognize and respect nature as a "speaking" subject leads us to question the common listening typologies circulating in academic and everyday discourse, but also the status of nature as only an object. Other applications of this extension are easily identifiable. We recognize the reverence given to sites such as battlefields, the World Trade Center location in New York, or to religious structures where many acts of human speech and action are ethical violations. It is silence, our ability to "be" in the place, which is the expected form of communicative engagement. These sites, with their animate and inanimate entities, must "speak" to our "listening" spirit. We must be co-present by engaging in a nonlinguistic discourse with the site that reaches the senses and/or emotions psychologically, and in so doing, aids us in learning about our relationship to this entity.

My objective finds precedent in the work of Cox (2007), who argues that it is the duty of scholars (and non-academics) to interrogate the human relationship with and ethics toward nature as mediated through communication practices and ideologies such as listening. The extension of dialogic listening practices to environmental contexts acknowledges that nature communicates, but does so in ways that may not conform to popularly accepted communication ideologies or definitions of "speech" (Carbaugh, 1999). It therefore challenges our typologies of listening. Recognizing that both inside and outside of academia listening and speech are socially constructed concepts (i.e., terms reflective of cultural ideologies) that are open to modification expands the possibility that the nature–human relationship, as reflected in Olson's writing, can change because of the understanding that comes through listening to nature as Other.

My effort adds to scholarship with a similar purpose. Oravec (1981) studied how the literary style of John Muir highlighted the sublime aesthetics of a sensory experience in nature as well as his disgust with the way in which humans were damaging a pristine wilderness in the name of civilization. Muir inspired in his admirers an ethical obligation to support preservationist policies that were instrumental in the formation of the national park system. Carbaugh (1999) examined the relationship of the Blackfeet people (Montana) with nature and how their culturally based listening behaviors situate a person in a shared scene with the Other where spiritual, historical, and natural elements "speak" and the listener's ethical obligations are inspired. Carbaugh writes: "One 'listens' to that immediately real, historically transmitted, spiritually infused, deeply interconnected world, to that complex arrangement in order to better understand that of which one is inevitably a small part" (p. 260). Finally, Milstein (2008) explores wildlife tourism (e.g. whale watching) as a setting where human silence during the tourist act opens a door where the whale as representative of nature can "speak" to a "listening" tourist. It opens the viewing tourist to a moment of sensory experience that is the only available option for dialogic listening and engagement with the Other.

Drawing on the rhetorical criticism technique of homological analysis (e.g. Brummett, 2004; McGee, 1980; Olson, 2002), this chapter identifies a distinctive description of a comprehension-based, human-centered listening practice (i.e., dialogic listening) in communication scholarship that can be extended to the experiences that function as communication in Olson's writings. As with the earlier examples of scholarship, the understanding that develops with the extension I am suggesting occurs not when we are projecting ourselves onto nature, but when we are listening ethically instead. Olson's work reveals that among the many sensory experiences that help us to understand the nonhuman "Other" that we may define as "wilderness," listening may be the most important because of how it positions us in relation to the nonhuman in a way that rejects anthropomorphism.

There is a twofold reason why this positioning of humans in relation to wilderness or nature outside of anthropomorphism is important in a book on "voice." First, the recognition of this voice(s), with the related implication of an entity with voice having value inherent in itself that was not granted or bestowed by humans, implies with such recognition an ethical obligation to listen to the entity with that voice. The two cannot be separated. Such an obligation, when fully actualized, could radically change the human relationship to the wilderness Other that has resulted in so much neglect. Second, the positioning is a radical shift in thinking about listening, which has historically situated listening to nature in appreciative terms. Listening to wilderness as Other, much like one would listen to another valued human, is uncommon in Western conceptualizations of this communication practice. The shift suggested in this chapter could lead to further rethinking of a listening typology and wilderness's place(s) within that typology.

Homology and ideograph

A rhetorical homology is a communication characteristic common across what should otherwise be unrelated artifacts or practices (Brummett, 2004; McGee, 1980; Olson, 2002). The characteristics may occur in the context inspiring a message, the channel by which a message is shared, and/or the patterns of experience or content in the message. This commonality is significant enough to the essence of each artifact to in some way define the character of both artifacts, yet not so foundational as to legitimate their pairing into a single genre. The commonality is not the product of universal human thought patterns, but instead represents the influence of social and (co-)cultural formations developed for sense-making processes within a community of individuals.

Brummett (2004) warns that the existence of these commonalities must be thoughtfully examined, being aware that "the nature of the connections among formally similar experiences is arguable but not absolutely, transcendently provable" (p. 11). The significance of the commonalities must be supported through reasoned argumentation and analysis by the observer. McGee (1980) adds that these commonalities exist because of strategic language choices and philosophic constructions present for the communicators generating or defining the individual artifacts or practices. "Human beings are 'conditioned,' not directly to belief and behavior, but to a vocabulary of concepts that function as guides, warrants, reasons, or excuses for behavior and belief" (p. 6). The use of this vocabulary by communicators may function for them as their argument and may represent or be the product of an ideology an audience will understand and accept. Thus, the most meaningful vocabulary revealed in the recognition of a homology "may be thought of as 'ideographs,' for...they signify and 'contain' a unique ideological commitment; further, they presumptuously suggest that each member of a community will see as a gestalt every complex nuance in them" (pp. 6–7). In other words, the homology reveals not only the presence of an ideology across the two artifacts or practices, but the manner in which the ideology legitimates, encourages, and/or constrains an audience's actions.

Scholars such as Brummett and McGee link homological analysis to the rhetorician Kenneth Burke, arguing that the commonalities identified are the equivalent of what he described as "God terms" (preferred frames of reference) and represent, in perhaps a single context, channel, or experience, a type of ideological argument about the proper order of things (behaviors, attitudes, social structures, etc.). McGee declares, "The important fact about ideographs is that they exist in real discourse, functioning clearly and evidently as agents of political consciousness" (p. 7). For this reason, it is possible to identify ideographs across several unrelated communication artifacts or practices. However, while ideological, "it is also true that there are special interests...separated one from the other precisely by disagreement regarding the identity, legitimacy, or definition of the ideographs" and what they should represent as a practice (p. 8). Hence, commonalities are not some indication of a universal truth, but are instead an indication of a (purposeful) social or cultural construction that can be challenged and altered/rejected by those who observe them.

Dialogic listening

I argue that two essays published in the *International Journal of Listening* explicate a portion of the listening as comprehension typology by describing dialogic listening practices that show an appreciation for the unique "Other-ness" of humans. Their explications establish a basis for understanding the dialogic listening (i.e. the homology) also present in Olson's writings. In their explication, Lipari (2009) examines the *moral obligation* to listen as a part of a dialogic ethics of communication while Tompkins (2009) examines how listening fosters a *moral sensitivity* toward human Others.

First, both authors contend that dialogic listening is an ethical obligation. Lipari maintains the obligation arises from a necessary need to understand an Other, what she refers to as "listening otherwise" (2009, p. 45). It requires us to embrace those aspects of an Other that are unfamiliar, removing us from the relative safety of understanding and connecting only through our similarities. Listening otherwise "subordinate[s] speaking to a kind of listening that speaks—a listening that is awakened and attuned to the sounds of *difference* rather than to the sounds of *sameness*" [emphasis original] (p. 45). Tompkins maintains that our obligation, or "moral sensitivity," occurs when we become concerned about the interests of others and employ a "moral imagination" that can make sense of ethical issues from their perspective (2009, p. 61). The listener who is unable or unwilling to engage in such mindfulness is morally insensitive, decreasing their overall ability to understand the interests or needs of an Other.

Second, both note that ethical listening benefits all actors in a communication exchange. Tompkins reminds us that how we conceptualize each part of the communication process (e.g., encoding a message, selecting a channel, interpreting a message) has ethical implications. Further, actors co-create meanings for the terms they employ, the environments they occupy, and their relationships; the implications of these co-creations multiply as the different meanings interact. "Ethical issues may become more significant when communicators do not recognize their relational connections with Others" or how their existences interact (2009, p. 63). In other words, we are selfish in the meaning-making process and do not try to understand an issue from any other position but our own. Correspondingly, Lipari declares that dialogic listening reduces the suffering of others. It constitutes a culture of caring, recognizing others as having an existence worthy of value and attention and, as a practice, creates a positive climate between the actors.

Third, both recognize "speaking"-centered biases present in the study of listening. Lipari recognizes the bias when she notes that interpersonal scholarship on dialogic ethics typically ignores the listener not just as an after-the-act response to a message, but also as a before-the-act inspiration to the act of speech. She argues:

> For in order to respond I must first listen—that is attend, observe, attune—and in doing so receive the otherness of the other. Without this listening there simply can be no response. There may be speaking, and there may be acting, but there can be no genuinely engaged response.... I may pretend to listen, or, even worse, pretend not to understand.... Conversely, I can refuse to listen by claiming to understand and yet be so immersed in the limitations and presuppositions of my own prior knowledge that I fail to listen altogether. I am struck

deaf and blind, in this sense, by what I *already know* and understand [emphasis original].

(2009, p. 47)

Clearly, the suggestion is that without listening, one can never understand an Other on their terms, but only on the terms one is willing to apply to the Other. Moreover, one can never understand what it is one should say in response to the Other.

Finally, both contend that listening is constitutive; the listening ideographs we adopt shape our relationship to the Other. Therefore, Lipari structures her challenge to the typical perspective around the concept of "compassion." As a way of knowing the Other, compassion based in empathy opens us to understanding by making sense of the Other's emotions and experiences. While admitting that one may not reach full understanding of the Other through compassion, either because certain experiences are beyond our comprehension or because of cognitive dissonance, failure to try to reach leads to a roadblock to any meaningful ways of knowing. The key for Lipari is to recognize the suffering the Other may experience and realize our own vulnerability to suffering.

Tompkins grounds her challenge in the constitutive nature of acts of communication, noting that when a person is aware that it takes two actors to communicate, attention is directed toward an Other who is affected by one's own communicative acts. A moral person would be sensitive to that impact, acting ethically to minimize harm. Because we enter into a cycle of communication acts, with no real beginning or end, we are part of a "network of relational connections" where "communicators have an opportunity to imagine seen and unseen Others involved in the communication process, and consider how [each] may be harmed or helped by an act of speaking or a decision to remain silent" (2009, p. 63).

These two approaches, which ultimately redefine and expand on what has previously been a restrictive human-to-human listening ideograph, remind us that the world we occupy is co-constructed and that any self-centered focus on our comprehension-based listening behaviors blind us to the needs, desires, and interests of an Other. One must listen to or through the channel and context the Other may choose or may be able to access. Included are "rhetorical and narrative conventions, language practices, and symbolic structures that create rhetorical presence or absence, and which influence understanding of an event, issue, or problem" (Tompkins, 2009, p. 69); in other words, the practices of meaning-making the Other may be able to employ.

Olson's wilderness Other

For Olson, coming to understand "wilderness," the Other that dominates his writing, is a psychological experience informed by the senses. Even though he earned a master's degree in zoology, suggesting that he could have focused his attention elsewhere, his writing emphasizes the senses. He describes the taste of a freshly caught trout cooked over his campfire, the smell of pine trees and decaying leaves, the feel of the warm water of a lake on his skin, and the visual splendor of the trout and the lake it came from. However, more than any other sensory experience, he highlights listening. Yet, it is far more than an appreciative experience that he highlights. It is, along with the support of the other senses, how he comes to "know" wilderness as an Other and how he argues his readers can come to the same knowledge. He reflects the centrality of nature-to-human listening as an ethical obligation in the name given to the cabin retreat he and his family occupied outside his hometown of Ely, Minnesota—Listening Point.

In *Runes of the North* (1963), he admits that his interest in listening ethically was inspired by the writing of poet Robert W. Service. He quotes "The Call of the Wild," where Service noted that humanity could listen to the wilderness Other. "There is a whisper on the night wind, there's a/star agleam to guide us,/And the wild is calling, calling...let us go" (p. 232). He references "The Law of the Yukon" to note that wilderness is listening in return. "I am the land that listens, I am the land that broods" (p. 235). For Olson, the "calling" (i.e., speaking) of wilderness is much more than a simple longing to be in the outdoors. Interpreting it as such would be a simplistic description of his work. It is a heartfelt sensitivity to the "voice" of the wilderness Other whose appreciation is not unique to Olson. Aldo Leopold in *A Sand County Almanac* (1949) and Rachel Carson in *Silent Spring* (1962) drew this same attention to listening. For example, in the opening chapter of her book, "A fable for tomorrow," Carson highlights a growing awareness of a loss of sounds from birds (e.g., "It was a spring without voices") and from bees ("The apple trees were coming into bloom, but no bees droned") as the result of chemicals and pesticides used in the environment (1962, p. 2). Her focus as a scientist and naturalist on listening emphasizes the recognition of an environmental problem first came, in part, by listening to sounds that were increasingly absent in the voice of nature.

For Leopold, we see this in chapters such as "The geese return," when he writes: "On April nights when it has become warm enough to sit outdoors, we love to listen to the proceedings of the [geese] convention in the marsh." His reason for listening is that the geese were addressing a "vehement controversy" and the last goose to honk had "the last word", with the "small-talk" now complete, how he wishes he could have been a muskrat on that pond so that he could have been closer and in a better position to understand (1949, p. 22). In "Thinking like a mountain," he writes about the howl of a wolf. He notes: "Every living thing pays heed to that call. [Despite what it might mean to deer, the coyote, the cowman, and the hunter]...there lies a deeper meaning, known only to the mountain itself. Only the mountain has lived long enough to listen objectively to the howl of the wolf" (1949, p. 129). It is in this chapter that he goes on to address his watching the green fade from the eyes of a wolf he and his travel companions had shot and killed. Finally, in "Song of the Gavilan," Leopold writes about the Rio Gavilan river and its "pleasant music," music that is "audible to every ear" yet can only be heard if "you live here for a long time" and can appreciate its message (1949, p. 149). While Carson and Leopold address listening as a side issue to their writing purpose, Olson makes it the centerpiece of his writing.

Just as Lipari and Tompkins argued that listening is an ethical obligation to the human Other, Olson argues we have this obligation to wilderness. In *Listening Point* (1958), a book paying tribute to his cabin retreat (in a chapter of the same title), he writes:

> Listening Point is a bare glaciated spit of rock in the Quetico-Superior country. Each time I have gone there I have found something new which has opened up great realms of thought and interest. For me it has been a point of discovery and, like all such places of departure, has assumed meaning far beyond the ordinary. (p. 3)

It is his listening to the wilderness Other that has opened him to new thoughts and discoveries. He finds that if we open ourselves to this Other, we learn that "nothing stands alone and everything, no matter how small, is part of a greater whole" (p. 4). Thus, "Listening Point is dedicated to recapturing this almost forgotten sense of wonder and learning from rocks and trees and all the life that is found there, truths that can encompass all" (p. 4). He claims that at such locations, one can "hear all that is worth listening for" (p. 7) and urges his readers to find their own special places to listen when he declares:

> I named this place Listening Point because only when one comes to listen, only when one is aware and still, can things be seen and heard.

> Everyone has a listening-point somewhere. It does not have to be in the north or close to the wilderness, but some place of quiet where the universe can be contemplated with awe. (p. 8)

The ethical move away from a self-centered appreciative ideology of listening and toward an expanded one of comprehension is clear and it mirrors the arguments of Tompkins and Lipari. It should be noted that part of Olson's love for this region is tied to recreating the listening experiences that Native Americans, voyageurs, and fur trappers had, experiences of understanding through portaging a canoe, building a fire, or sleeping under the stars. These experiences, he admits, are practically lost to his generation but were what connected earlier generations to the wilderness Other.

Listening must benefit all actors, a point also made by Tompkins and Lipari. When Olson cited Service, he recognized that listening to wilderness meant that wilderness listened in return. The growing human failure to listen in return was an ethical injustice to the Other, stripping it of the opportunity to be heard and to become a fully developed "self." In "Flying In," from *The Singing Wilderness* (1956), Olson takes a plane ride to a hard-to-reach lake. The plane, which landed on the water, dropped him at a favorite campsite and left. He writes: "This is what I had dreamed of doing, but for the first time in my life I had failed to work for the joy of knowing the wilderness; had not given it a chance to become part of me" (p. 115). Because he had not canoed in, he had not had a sensory experience or understanding-filled exchange with his surroundings. It is not just that he missed the sounds associated with canoe travel, but an entire relationship was displaced by a mode of transportation that silenced the voice of wilderness and silenced his response. This experience later led to his work toward the passage of laws banning motorboats and plane landings in the Boundary Waters Canoe Area and Voyageurs National Park (which occupy a large area in northern Minnesota).

Olson explores the human benefits to listening to the wilderness Other in the prologue to *Wilderness Days* (1972). While arguing that ancient humans, Native Americans, and voyageurs were more in touch with wilderness than most modern humans are, the possibility of re-establishing an understanding of the wilderness Other still exists. He writes:

> There are moments of insight when ancient truths stand out more vividly and one senses again his relationship to the earth and all

life. Such times are worth waiting for, and when they come in some unheralded instant of knowing are the purest gold. (p. xx)

Though modern culture has distanced humanity from and perhaps eliminated *Wild*-erness as a location untouched by industrialization, Olson maintains that the fundamental way of knowing the "wilderness" that still exists is deep in our being. He characterizes returning to that way of knowing when he writes: "It is a communion with the natural world, a basic awareness of earth wisdom which, since the beginning of man's rise from the primitive, has nourished his visions and dreams" (p. xx).

Although it is difficult to locate specific examples where Olson reflects upon wilderness as the listening Other or describes fully the benefits it is receiving, it is possible to find such acknowledgement just below the surface of his work. It should not be surprising that he would have limited references to this theme. During his time as a writer, such sensitivity would initially have been rare and not popular enough to garner the type of following he would eventually build. Backes, in his edited collection of Olson's articles and speeches (see Olson, 1997), positions a young Olson as a mediocre fiction writer and preservationist. He was trying to incorporate the (listening for the purposes of) wilderness-protecting thesis that characterizes his stories published in what were often conservationist-leaning magazines that more broadly situated wilderness in terms of managed use of resources versus protected spaces. Likewise, the hunting and fishing magazines and newspapers he targeted for his writing emphasized the theme of the successful outing—and the techniques that led to that success. This context created an important tension (and source of criticism) for his work. He ultimately wrote for wilderness protection, but to do so he had to write about and encourage people to appreciate wilderness by entering into the very spaces he wanted to protect from overuse and abuse.

Further, arguing for the personal (i.e., non-materialistic, psychological) benefits of a wilderness experience was a widely accepted tradition across the various forms of the nature-writing fiction genre. This is especially true for many of the literary figures Olson read (e.g., Wordsworth, Coleridge, and Whitman). Yet, Olson positioned his particular approach to that psychological benefit as an attachment and need for listening to wilderness that was a product of our historic memory as members of the human race who historically had a much stronger understanding of wilderness in earlier times. Olson's approach of blending emotions and sensory experiences into a romanticism was less common (and thus less accepted) in the magazines and newspapers where he would get his start as a writer.

However, in *Songs of the North* (1987), an edited book of his work, in a chapter titled "Song of the North," an example does exist:

> Someone said, "Do not take away from any man his song," and when I think of the one I have heard, what it has meant and how it has guided me in whatever I have done, I know this warning is true, for when a song is lost, it is seldom replaced. (p. 70)

Any voice silenced, including humanity's for nature to listen to, is a song lost. Wilderness listened to Native Americans and voyageurs, in Olson's examples, when each portaged the divides between the many lakes of northern Minnesota with awe and reverence. If contemporary wilderness does benefit from listening, it is perhaps not in some reaction of wilderness to what humans have spoken but in the sensitivity humans will acquire when interacting (i.e., recreation) in wilderness, a sensitivity Olson found lacking whenever he came across a remote campsite littered with garbage or treated with insensitivity to its fragile state (e.g., improper fire rings).

The third concern of Lipari and Tompkins was that the practice of human-to-human listening (as well as the ideologies that sustain that practice), as a part of the communication process, has gone relatively unexamined; a point Olson finds in the approach to wilderness as well. In the prologue to *Wilderness Days*, reflecting on Listening Point, he notes that listening was a common part of primitive interactions with our natural surroundings; but, as already noted, this experience is largely lost today:

> I know this concept is one of the greatest satisfactions of man, that when he gazed upon the earth and sky with wonder, sensed the first vague glimmerings of meaning in the universe, the world of knowledge and spirit was opened to him. I also believe that, while such inherent joys are often lost, deep within us their latent glow can be fanned to flame again by awareness and an open mind.
>
> (1972, p. xxii).

Olson concludes this passage by contending that nature-to-human listening need not be isolated to areas far-removed from cities; that any place that is quiet and allows one to hear nature-made sounds—even a backyard garden in the middle of the city—can serve as a listening point.

In *The Singing Wilderness*, he also reflects on this nearly lost art. He characterizes it as being "concerned with the simple joys, the timelessness and perspective found in a way of life that is close to the past" (1956, p. 5). While it might be easy to interpret Olson's concern for the past as nostalgia for the time of the Native Americans or the voyageurs, his timelessness and perspective is more concerned with reconnecting individuals with a lost knowledge of the Other. He suggests that listening "seems to be part of the hunger that all of us have for a time when we were closer to lakes and rivers, to mountains and meadows and forests, than we are today" (p. 6). One must be reflective of what one listens to, because it is "sometimes not until long afterward when, like an echo out of the past, you know [the message wilderness was speaking] was there in some quiet place or when you were doing some simple thing in the out-of-doors" (p. 6).

The final concern of Tompkins and Lipari was the role listening, and the communication that comes from it, plays in constituting our shared world (and not just the understanding of listening itself). In "Silence," a narrative from *The Singing Wilderness*, he writes about how, after listening, his world has been constituted anew:

> The sun was trembling now on the edge of the ridge. It was alive, almost fluid and pulsating, and as I watched it sink I thought that I could feel the earth turning from it, actually feel its rotation. Over all was the silence of the wilderness, that sense of oneness which comes only when there are no distracting sights or sounds, when we listen with inward ears and see with inward eyes, when we feel and are aware with our entire beings rather than our senses. I thought as I sat there of the ancient admonition "Be still and know that I am God," and knew that without stillness there can be no knowing, without divorcement from the outside influences man cannot know what spirit means.
>
> (1956, pp. 130–131)

The paragraph is one of Olson's clearest indications of how, after listening to the narratives wilderness has shared with him, his shared world with wilderness has been reconstituted and he now has established a frame for understanding the Other and himself. The silence, away from industrial life, is important enough that he advocates for its protection: "How often we speak of the great silences of the wilderness and of the importance of preserving them and the wonder and peace to be found there!" (1956, p. 134).

Olson does not anthropomorphize the wilderness Other in such a way that we simply cross-apply listening behaviors; to do so would disrespect the very nature of the wilderness Other. The sounds wilderness offers are not so much "voice," as he (and the earlier Leopold example) notes, but music. In "Wilderness Music," from *The Singing Wilderness*, he points to wilderness' musical qualities:

> There are many types of music, each one different from the rest: a pack of coyotes and the wild, beautiful sound of them as they tune up under the moon; the song of a white-throated sparrow, its one clear note so closely associated with trout streams that whenever I hear one, I see a sunset-tinted pool and feel the water around my boots. The groaning and cracking of forming ice on the lakes, the swish of skis or snowshoes in dry snow—wilderness music, all of it, music for Indians and for those who have ears to hear.
>
> (1956, p. 203)

The wilderness music, one form of "voice" for wilderness, reflects the "speech" nature is able to make toward constituting its character and the character of our relationship to it. In an earlier paragraph, it was noted that being in communion and belonging is key to the listening-based orientation Olson takes toward the wilderness Other. This, I will argue in the conclusion to this essay, draws together what Lipari and Tompkins also argued in their approaches to human–human based listening as compassionate, moral, and constitutive of relationships.

The *Wilder*-ness of listening to the voice of wilderness

In this section, I explore the implications of extending dialogic listening behaviors to a context as familiar as the environmental behaviors that Olson once encouraged and which many readers of this book may be interested in furthering. One place to begin is with the practices of environmental social movement actors. As a justification for building this extension here, previous research has employed comparable analysis of social movements (e.g. Black, 2003; Olson, 2002). Black (2003) used homological analysis to examine rhetoric giving voice (thus demanding listening) to human fetuses in right-to-life debates and to animals in animal-rights debates. Both grant consciousness to a "life" whose "status" is unlike that of their defenders. Alternatively, Olson (2002) studied the homological similarities in how voice (and therefore listening) is denied in the context of sport hunting, hate crimes, and rape. Studies such as these uncover commonalities in language choices, images, and/or outcomes that reflect strategic attempts at generating behavioral responses (i.e. degrees/types of listening) in the audiences to the respective social movements associated with each issue. Moreover, both situate the animal as Other.

While neither human-to-human nor nature-to-human listening is itself the foundation of a "social movement," the listening behaviors Olson associates with environmental awareness and sensitivity have been and remain a significant feature of the advocacy occurring in many environmental justice and other social movements (e.g., Anguiano, Milstein, De Larkin, Chen, & Sandoval, 2012; Dreher, 2009; Milstein, Anguiano, Sandoval, Chen, & Dickinson, 2011). Therefore, ethical listening behaviors should be extended and/or expanded on to include not only the listening to nature as a stakeholder/Other that this chapter examines and encourages, but also the dialogic listening to human Others who are stakeholders in any environmental justice debate.

However, dialogic listening cannot be the entire basis of an environmental revolution; therefore, what can we reasonably take away from this analysis of Olson's extensive nature writing that will aid current environmental social movements? Watts (2001) suggests part of the answer. In asking if "voice" is "an original impulse of being," he maintains that scholars and non-scholars alike already use the term "voice," extending it to any number of linguistic and nonlinguistic acts which demand ethical obligations for the rhetor and the receiver. Furthermore, " 'voice[...]' announces the body's presence" (Watts, 2001, p. 180) and "must be considered an ontological concern, not a physical one" (p. 181). For Watts, "voice" is the essence of the Other in a shared location, a signification with purpose. An objective for social movement advocates, then, should be opening up the possibilities of voice for those (including the natural realm) on the margins of society. To "hear" this voice and be able to act ethically with this relational partner, in the spirit of Lipari's earlier argument, one must first listen. Watts asserts:

> "Voice" is a particular kind of speech phenomenon that pronounces the ethical problems and obligations incumbent in community building and arouses in persons and groups the frustrations, sufferings, and joys of such commitments. Rhetorical "voice" is not a unitary *Thing* that inhabits texts or persons either singularly or collectively. It is itself a *Happening* that is invigorated by a public

awareness of the ethical and emotional concerns of discourse [italic original].

(2001, p. 185)

As Olson, Tompkins, and Lipari declare with their statements on listening, voice is the recognition of the Other acknowledged through the ethical obligation of listening. To leave "voice" as an element of communication style in the strict rhetorical sense (i.e., belonging only to human-to-human interactions) is to leave it far too meaningless (and is to leave some voices unheard when they do not conform to an accepted cultural or ontological standard).

Thus, we must take away from Watts, Olson, Tompkins, and Lipari the realization that if voice is the ontological representation of being, of a form of existence recognized and having any value as a subject/Other, then dialogic listening is the only way that the ontological fulfillment of that being can occur (for humans or for nature). Consequently, the ethical problems and obligations associated with recognizing and appreciating the "voice" of "wilderness" are resolved through two conditions. First, by being better dialogic listeners to the wilderness Other and the manner/form in which it expresses its "voice." Second, by being more thoughtful about and/or challenging where necessary the ideologies (as they are manifest in any typology) that define or justify "listening" because one cannot achieve the first condition on a meaningful societal scale without also addressing the second condition. Such an effort will not be easy; however, it may open humans to the suffering of the wilderness Other, create an environment of care and compassion for "wilderness," and constitute appropriate behaviors moving forward. All of these efforts are well within the scope of Olson's writings. Failure to listen, if Watts is correct, leaves little opportunity for "wilderness" of achieving subjecthood.

Tuan (1993) acknowledges that when something falls silent, we conceptualize it as being "dead." Silence strips an entity of a particular kind of presence, value, and worthiness of respect. Silence may come not just when something is still or quiet, but when one ignores or no longer appreciates this entity's "voice." Historically, human communities have silenced those whose voice they wanted to disappear (e.g., minorities, children, and the poor). The injuriousness of the silencing may have come through a thoughtless act of ignoring the Other or a strategic act that stripped voice away. Just as humans silence other humans, wilderness is silenced by any practice or philosophy that denies it a figurative "voice" and an ethics of presence, value, and respect.

In this same spirit, Nash (2001) contends that it may be because of the buzz of modernity that one can come to appreciate the "voice" of wilderness. Citing Olson's *Listening Point*, he argues that a primary way to appreciate nature's voice is by recognizing its absence in areas we already occupy (e.g. industrial areas), being overwhelmed by its opposite in our mass culture (e.g. television), and its temporary silencing where normally its presence would be obvious (when a passing car drowns the sounds of nature). Just as humans may appreciate the benefits that interactions with other humans entail—especially when one has been distanced from this interaction, so too does appreciation for nature's voice increase when juxtaposed against its absence. Both Olson and Tuan identified this voice as a spiritual sound of the heavenly spheres, of a wilderness with an ability to heal the human spirit.

Part of the answer to the earlier question of what we can reasonably take away from this analysis that will aid current environmental social movements is also found in our broader contemplation of Other-directed ethics. Just as dialogic listening traditionally reflects a particular ethical concern for the human Other, so too does listening to wilderness reflect a particular environmental ethic. This chapter is, without question, adopting an extensionist approach from dialogic ethics to environmental ethics. Sandler (2006) argues the extensionist "approach begins with a character trait considered to be a virtue in interpersonal interactions or relationships and proceeds by arguing that the virtue ought to be operative in environmental interactions or relationships as well" (p. 248). Sticking with what is virtuous is, according to Des Jardins (2001), preferable because it is easier to argue that humans have an ethical obligation to protect wilderness than it is to argue that wilderness has "rights." One virtuous way to protect wilderness is to show respect to its "voice" or narrative.

So, one may ask, what is this "voice" and how does one "hear" it? Olson tells us that his physical recognition of it came to him through his senses. It was the smell of smoke from a campfire or the sound of a loon calling from across a lake. However, Olson also suggests that "hearing" these things required us to "be" in the place. This means that our meaningful cognitive recognition of and response to what we have sensed, our "presence" in the relationship, must come from an Other-centered approach where our attention is on the "source" of what we have sensed (and not simply stopping with our physical enjoyment of the sensation itself). In other words, it is not appreciative listening. It means that we are willing to contemplate the "meaning" or purpose of what we have sensed to the natural "Other" that is the source of what we have just sensed. In doing so, we open ourselves to—if not giving intentionality toward—the Other's "presence." All of this is counter to giving complete attention to our own needs and desires in our relationship to a place as "Other."

Callicott (1984) defines this virtuous sentimentality as "biophilia." He asserts that this is an ethic, grounded in Darwin, Leopold, and Hume, which provides for an intrinsic value for the nonhuman Other but, much like the example of the Blackfeet noted at the start of this chapter, is humanistic in the extent to which it requires humans to express or be the measure of the values. It defines an entire biosphere as a community where every entity is a neighbor deserving of ethical consideration. Under a similar approach to the ethic, Saito (1998) contends that an appreciation for wilderness must move beyond modernist approaches that portray it as a resource or as backdrop to human narratives and, instead, acknowledge that nature has its own design or purpose (i.e., "voice/presence"). That purpose is understood through sensory-based approaches that appreciate wilderness aesthetics as an inroad to generating a moral sensibility toward wilderness. Nature writing and travelogues such as Olson's often focus on sensory experiences that tell nature's story—the taste of a trout, the smell of the pine, the sound of a waterfall, and so on. These narratives are significant for the way they can make us sensitive to all relationships with the Other if we only listen (or use whichever appropriate sense) to acknowledge and appreciate these "voices."

The listening ideograph (i.e., the dominant typology and my revision of it through the homological analysis) is, as Burke (1968) contends, "equipment for living." In the context of this chapter, its revision or extension reflects a strategic choice on the part of Olson to construct an ethic that would influence his audience to a desired act— environmentally responsible behaviors. Olson's particular use of the listening ideograph (or its current potential for environmental awareness) may not have been realized were it not for the type of homological analysis done in this chapter. Yet, as Olson (2002) notes and I readily admit, audience recognition and/or understanding of the substance of an ideograph does not necessarily guarantee changes in action— in this case, better environmental behaviors because of the listening employed. Environmentally responsible ethics are far too complex for such a simple fix.

Finally, instead of rehashing historic arguments about why the ethics I am advocating are important, with the space remaining I will ground these dialogic ethics in an application to the environmental justice movement. DeLuca (2007) contends that environmental justice focused on wilderness protection has been under attack by those advocating a human needs-based emphasis (e.g., clean air and water). The claim is that human needs must come before any concern for nature. Any failure to address these needs can be reduced to racism, classism, or other discriminatory practices that leave particular human communities threatened. While a different set of claims is employed by industrialist voices who object to efforts to protect wilderness resources, the outcome of both sets of claims is much the same for wilderness. Yet, as DeLuca maintains, giving up on wilderness in an effort to protect human communities or the economy is a choice made in error; the needs are intertwined.

Initially, we must accept that the "wilderness" that may be saved through listening may not be the one reflected through the writing of Olson. DeLuca describes this as a "white wilderness" because the individuals who rhetorically constituted its existence were white individuals—usually elite men. He notes, "the values of elite 'white' culture were inscribed in a vision of pristine, sublime wilderness that subsequently became a foundational value of the preservation movement" (2007, p. 38). Yet, DeLuca finds it concerning that criticism of this construct can so easily lead some environmental justice groups to abandon efforts to protect "wilderness" by any definition. Postmodernism, he notes, should open our mind and senses to new understandings of wilderness as a site humans occupy, just as it is opening our understanding of the human Other. Olson, while certainly informed by and writing within a language and style that reflected his white, elite status, may have experienced and interacted with a wilderness Other that was not completely defined by his ethnicity or class. In fact, the wilderness he writes about shares much with what the Blackfeet people noted earlier. Put another way, it is not Olson's wilderness that we (or he, for that matter) would strive to protect, but the wilderness "Others" many of us have come to know and we are reminded of when reading a particular author like Olson, Muir, or Thoreau.

Listening in the human-focused side of the environmental justice movement, like much social movement rhetoric, is slowly making privileged groups aware of the conditions under which less privileged individuals are living. We are acknowledging the "voice" and "presence" of these groups. It is this very type of listening that Tompkins and Lipari advocate. Currently, both the privileged and the unprivileged human groups may have very little interest in nonhuman wilderness ethics or may not see the link between human-to-human and nature-to-human dialogic listening. Yet, the attention drawn to human living conditions implicates the real possibilities of nature as Other-directed listening into an all-encompassing ethic of environmental justice.

If DeLuca is correct that environmental justice should ultimately be environment-oriented in a way that both promotes and protects what is best for our entire biosphere, then listening to wilderness is as important to environmental justice as listening to the stories of the injured and oppressed who happen to be human (see Cronon, 1996; Sterba, 2005). Listening to humans who have been hurt reduces the ontological separation between the injured and the injurer, between our industrial and consumerist practices and their effects on both our human and nonhuman neighbors. Likewise, basing our understanding of wilderness on the inaccurate ontological and epistemological constructions of our past denies wilderness existence and a voice in contributing to the shaping of its and our future in the biosphere. Furthermore, limiting our understanding of wilderness to the previously oppressive constructions of modernism leaves silent places in the biosphere that have either been under-appreciated or that have been radically altered by the presence of humanity and yet could still be wilderness—even in Olson's conceptualization.

As I have noted in this essay, "listening" and "voice" will still be human constructs—for the pristine and the polluted-on natural Other. Therefore, agency, in the voice of wilderness, can only be expressed through narratives that we mentally construct or which take the form of essays like that of Olson. Thus, authors like Olson can be a guide and resource for building in us and our community the "biophilia" ethic Callicott advocates. In science, economics, and literature classrooms (e.g. Williams, 2002; Klenk, 2008), for example, the writings of these individuals could be added to a collection of "essential" authors. As Watts notes,

> As we all increasingly feel displaced from one another, occurrences of "voice" represent events in which we can characterize our commitments and sentiments toward social spaces. "Voice" grounds us by reminding us of our situatedness. We are also reminded of our own needs. Such an interest for criticism serves to return us to an examination of the basic requirements for living with ourselves and with others."
>
> (2001, p. 192)

Likewise, as Milstein (2008) noted, a context such as wildlife tourism opens us to opportunities where nature can "speak" to a listening Other.

However, the opportunities to listen do not have to occur only in the presence of the most impressive of wildlife or at the most grand of national parks. Olson, Carbaugh, and others have noted that listening can happen on a much more localized and personal scale. If one comes to accept that the nature in one's backyard or a city park "speaks" and that one can "listen" with an ethical obligation, a relationship grows.

Finally, a chapter like this is necessary because many cultural frameworks in academia define nature directly or indirectly as "separate, other, and mute, allowing for and in some ways encouraging environmental practices that are exploitive and destructive" (Milstein, 2008, p. 187). "Listening" to the Other that is wilderness is essential for an ethic of environmental responsibility to germinate and grow in academia as well. To this end, this chapter extends the great work done by scholars of human-to-human listening to a reasonable nature-to-human context. Listening is an ethical orientation toward the wilderness Other as well as a way of knowing that fits the criteria for an ethical obligation established by King (2006), who argues: "In order to critique dysfunctional belief systems, we must find a combination of strategies that will both articulate new beliefs and yet also embody them in sustainable, supportive practices" (p. 175). Conceptualizing listening in the way explored in this chapter allows not only for critique of our existing interrelationships, but also for the expression of alternative beliefs that could lead to more sustainable practices. As the environmental justice movement illustrates, this may be an approach fitting all acts of environmentalism.

References

Anguiano, C., Milstein, T., De Larkin, I., Chen, Y., & Sandoval, J. (2012). Connecting community voices: Using a Latino/a critical race theory lens on environmental justice advocacy. *Journal of International and Intercultural Communication, 5*, 124–143.
Backes, D. (1997). *A Wilderness Within: The Life of Sigurd Olson*. Minneapolis: University of Minnesota Press.
Black, J. E. (2003). Extending the rights of personhood, voice, and life to sensate others: A homology of right to life and animals rights rhetoric. *Communication Quarterly, 51*, 312–331.
Brummett, B. (2004). *Rhetorical Homologies: Form, Culture, Experience*. Tuscaloosa: The University of Alabama Press.
Burke, K. (1968). *Counter-Statement*. Berkeley: University of California Press.
Callicott, J. B. (1984). Non-anthropocentric value theory and environmental ethics. *American Philosophic Quarterly, 21*, 299–309.
Carbaugh, D. (1999). "Just listen": "Listening" and landscape amongst the Blackfeet. *Western Journal of Communication, 63*, 250–270.
Carson. R. (1962). *Silent Spring*. Boston: Houghton Mifflin Company.

Cox, R. (2007). Nature's "crisis disciplines": Does environmental communication have an ethical duty? *Environmental Communication, 1*, 5–20.

Cronon, W. (1996). *Uncommon Ground: Rethinking the Human Place in Nature.* New York: W.W. Norton & Company.

DeLuca, K. (2007). A wilderness environmentalism manifesto: Contesting the infinite self-absorption of humans. In R. Sandler & P. C. Pezzullo (Eds), *Environmental Justice and Environmentalism: The Social Justice Challenge to the Environmental Movement* (pp. 27–56). Cambridge: The MIT Press.

Des Jardins, J. R. (2001). *Environmental Ethics: An Introduction to Environmental Philosophy.* Belmont, CA: Wadsworth.

Dreher, T. (2009). Listening across difference: Media and multiculturalism beyond the politics of voice. *Continuum: Journal of Media & Cultural Studies, 23*, 445–458.

King, R. J. H. (2006). Playing with boundaries: Critical reflections on strategies for an environmental culture and the promise of civic environmentalism. *Ethics, Place and Environment, 9*, 173–186.

Klenk, N. (2008). Listening to the birds: A pragmatic proposal for forestry. *Environmental Values, 17*, 331–351.

Leopold, A. (1949). *A Sand County Almanac.* New York: Oxford University Press.

Lipari, L. (2009). Listening otherwise: The voice of ethics. *The International Journal of Listening, 23*, 44–59.

McGee, M. C. (1980). The "ideograph": A link between rhetoric and ideology. *The Quarterly Journal of Speech, 66*, 1–16.

Milstein, T. (2008). When whales "speak for themselves": Communication as a mediating force in wildlife tourism. *Environmental Communication, 2*, 173–192.

Milstein, T., Anguiano, C., Sandoval, J., Chen, Y., & Dickinson, E. (2011). Communicating a "new" environmental vernacular: A sense of relations-in-place. *Communication Monographs, 78*, 486–510.

Nash, R. F. (2001). *Wilderness and the American Mind.* New Haven: Yale University Press.

Olson, K. M. (2002). Detecting a common interpretive framework for impersonal violence: The homology in participants' rhetoric on sport hunting, "hate crimes," and stranger rape. *Southern Communication Journal, 67*, 215–244.

Olson, S. F. (1956). *The Singing Wilderness.* New York: Alfred A. Knopf.

Olson, S. F. (1958). *Listening Point.* New York: Alfred A. Knopf.

Olson, S. F. (1963). *Runes of the North.* New York: Alfred A. Knopf.

Olson, S. F. (1972). *Wilderness Days.* New York: Alfred A. Knopf.

Olson, S. F. (1987). *Songs of the North: A Sigurd Olson Reader.* Ed. H. F. Mosher. New York: Penguin Books.

Olson, S. F. (2001). *The Meaning of Wilderness.* Ed. D. Backes. Minneapolis: University of Minnesota Press.

Oravec, C. (1981). John Muir, Yosemite, and the sublime response: A study in the rhetoric of preservationism. *Quarterly Journal of Speech, 67*, 245–258.

Purdy, M. (1991). What is listening? In D. Borisoff & M. Purdy (Eds.), *Listening in Everyday Life: A Personal and Professional Approach* (pp. 1–20). New York: Lanham.

Saito, Y. (1998). Appreciating nature on its own terms. *Environmental Ethics, 20*, 135–149.

Sandler, R. (2006). A theory of environmental virtue. *Environmental Ethics, 28*, 247–264.

Sterba, J. P. (2005). Global justice for humans or for all living beings and what difference it makes. *The Journal of Ethics, 9,* 283–300.
Tompkins, P. S. (2009). Rhetorical listening and moral sensitivity. *The International Journal of Listening, 23,* 60–79.
Tuan, Y. (1993). *Passing Strange and Wonderful: Aesthetics, Nature, and Culture.* New York: Kodansha International.
Williams, D. (2002). Reconnecting body and mind with Earth. In S. Fletcher (Ed.), *Philosophy of Education 2002: Proceedings for the annual meeting of the Philosophy of Education Society* (pp. 53–56). Richmond, VA: Virginia Commonwealth University.
Watts, E. K. (2001). "Voice" and "voicelessness" in rhetorical studies. *Quarterly Journal of Speech, 87,* 179–196.
Wolvin, A. D., & Coakley, C. G. (1993). A listening taxonomy. In A. D. Wolvin & C. G. Coakley (Eds.), *Perspectives on Listening* (pp. 15–22). Norwood, NJ: Ablex Publishing Corporation.

11
Listening to the Natural World: Ecopsychology of Listening from a Hawai'ian Spiritual Perspective

Yukari Kunisue

> *E Ho Mai*
> E Ho Mai ka ike mai luna mai e
> O na mea huna no'eau o na mele e
> E Ho Mai, E Ho Mai, E Ho Mai, e 'a
> Give us knowledge from above
> All the wisdom of the songs
> Give us, grant us, bestow us
>
> (traditional chant before
> *ho'oponopono*)

For tens of thousands of years the lives of Hawai'ians have coexisted with nature. The vast ocean and the volcanic islands have provided abundant gifts to the islanders and sustained every living being in intricate balance. As in many indigenous orally based cultures, people's livelihoods in Hawai'i have been deeply intertwined with nature. People worship nature in their daily lives with chants, prayers, and dance. It is not possible to discuss Hawai'ian life and its people, even their psychology and communication, without referring to the context of nature and what Abram calls the "larger more-than-human worlds" (Abram, 1995, p. 304) that expand beyond limited human consciousness.

In this essay, I am offering an ethnographic description of Hawai'ian communication in relation to listening to nature. My view is based on training as a scholar of transpersonal psychology, and this analysis explores how traditional Hawai'ian communication confirms a "transhuman dialogue" (Rogers, 1998). I became intrigued by modern life in Hawai'i, which manages to sustain the spirit of *ha* (wisdom) carrying *mana* (life force), and *huna* (the secret wisdom of life). I engaged in observation of daily life with local Big Islanders, where nature dominates human life (not the other way around as on Oahu Island in Hawai'i), with a hope of exploring how they live life deeply connected with the natural environment. I started listening to how people in Hawai'i view and use their perceptions, senses, and knowledge through seeing and hearing. My intention is to provide another tool for scholars in communication to understand voice in nature and beyond modern human consciousness.

Each day in Hawai'i, people in their uniquely modest way tell me their stories (Hawai'ians call it "talk stories"), and share mythologies and ancient legends. I hear chanting and *pule* (praying) on the lava beach on an early Sunday morning. I view authentic *hula* dances, both on stage at the internationally acclaimed Merry Monarch Festival, or simply practicing in the park. They look deeply awed and proud, carrying the heritage and being nature themselves. I observe Hawai'ian *kupuna* or elders greet each other by *honi ihu* (nose kissing). I learned an ancient mediation method, *ho'oponopono*, conducted by skillful elders. I see that all these rituals are a Hawai'ian way of listening to the voice of their natural environment.

This Hawai'ian way of living is precisely the stance of ecopsychology. Ecopsychology is a new discipline, yet it seeks and rediscovers ancient understanding of human interaction with the natural environment, as many indigenous traditions help us remember. Ecopsychology integrates ecology and psychology and, unlike mainstream psychology, ecopsychology recognizes humans as "actors on a planetary stage who shape and are shaped by the biospheric system" (Roszak, Gomes, & Kanner, 1995, p. 4). In the world of ecopsychology, as in the Hawai'ian cosmic sphere, humans are not independent units, manipulators, separated from their surroundings. Though modern psychologists have tried hard to establish an individual's boundary as being separate from others, humans never exist separated from their environment. For instance, Dudley (1990) states that in Hawai'ian *wai* the water of life and man are inseparable and inextricably one. They are both parts of "a continuum joined ultimately with all other living things—plants, animals, winds, mountains and the like" (Dudley, 1990, p. vii). We are a part of the ecology. Or, to be more precise, we are ecology *ourselves*.

For Hawai'ians, nature, just as the gods, is sentient and conscious of beings that could hear, talk, sense, breathe, and feel. Carbaugh (1999), in his exploration of listening in the Blackfeet tribe, expresses the strong link between "sacredness, place and the 'listening' form" (p. 258). The Hawaii'an way is not unlike Blackfeet tradition; in their "historical, tranquil and beautiful places, the land (islands in Hawaii's case) speaks" (p. 258). When ravens "spoke" to Carbaugh, Two Bears offered him Blackfeet wisdom about the message from nature: "This is for the listener to decide. The meaning will be the listener's" (p. 260). In Hawai'i, as well, each listener understands in his or her own way what nature is saying.

In Hawai'i, when you see a fisherman in the morning preparing for fishing, you should never ask him if he is actually going fishing. You could ask, using other euphemisms such as "going for a ride?" or "going for mountains?" but never directly about fishing. It is because, Hawai'ians believe, the fish could "hear" (Dudley, 1990, p. 1) the fisherman's intention and are certainly not going to wait around for him. By communicating with nature in such an intimate way, Hawai'ians are interrelating with nature as if it is as conscious and sentient as they are. A parallel can be drawn with the Nez Perce tribe's Weyekin Principle (Salvador & Clarke, 2011) that "emphasizes both a different way of listening to and a different way of speaking about the environment" (p. 245). Nez Perce people use the term "weyekin" to symbolize the close connection between humans and animals (such as wolves). It is not just a mystical or metaphysical relationship", but refers to "direct, corporeal, and material experiences with the non-humans" (p. 247).

As a modern ecopsychologist, Laura Sewall (1995) terms this capacity as "skillful ecological perception" (p. 201), indicating that ecological consciousness is a set of skills that anyone, even modern industrialized citizens like us, could acquire with a sincere effort. Our lower level self obtains the skills when the forgotten "ecological self" experiences permeability and fluidity of boundaries between self and beyond self. As Roszak et al. (1995) state, "ecopsychology proceeds from the assumption that at its deepest level the psyche remains sympathetically bonded to the Earth that mothered us into existence" (p. 5). As an emerging synthesis, ecopsychology teaches that what nature and humans feel is fundamentally the same.

Communication in the Hawai'ian world never departed from such an ecopsychological view. Communication could not possibly be limited to only that between humans, a Hawai'ian might say. It is a way to connect people to the surrounding environment including animals, plants, fish, rain, flowers, waterfalls, waves, and volcanoes. Hawai'ians worship natural phenomena in their daily life by offering respect and awe to hundreds of deities. Communication takes the form of wider perceptions and sensations, not limited to verbal or visual cues. The paradox that humans have in communication is that verbalizing about and of nature changes the very form of communication. In other words, as soon as humans speak about nature it "alienates humans from the material-physical reality of nonhunman nature" (Maxcy, 1994, p. 330). Humans learned to express gratitude and appreciation through prayer, dance, and chanting (Lewis, 1983). Hawai'ian language uses metaphors and underlying meanings so that its speakers can avoid such alienation.

Listening, too, is part of wider and more intelligent activities coming from within and beyond human capacity. Without knowledge of how to respect and maintain balance among all things that exist in the larger world, humans could not survive island life. Together with resources from land and ocean, wisdom pours into daily human life from all around the surroundings as long as she or he is listening deeply and attentively to nature's voice.

As a researcher of listening and scholar of transpersonal psychology, I became intrigued by modern life in Hawai'i, which manages to sustain the spirit of *ha* (wisdom) carrying *mana* (life force), and *huna* (the secret wisdom of life). I started listening to how people in Hawai'i view and use their perceptions, senses, and knowledge through seeing and hearing.

Ha: Divine breath and listening in Hawai'ian epistemology

As stated above, Hawai'ian language is symbolic, containing many metaphors and subtle meanings (Lewis, 1983). In an oral heritage, without a written form for thousands of years, people's communications were attuned to listening and observing. Listening in particular has never been a passive part of Hawai'ian language, as it is the crucial gateway to the sounds, tunes, both overt and subtle meanings, and abundant images provoked in the human mind and heart. In the traditional education or *a'o*, Hawai'ians set ground rules:

> *Nan. ka maka*: Eyes on teacher, environment and fellow learners
> *Ho'olohe ka pepeiao*: Listen and hear with all senses
> *Paa ka waha*: Close your mouth
> *Hana ka lima*: Work with hands
> *Ninau*: Finally, you may ask questions
>
> (Bezilla, 2012, September 27)

These simple rules tell how silence and attentiveness are crucial for Hawai'ian learning, which often originates from nature. The attentiveness here signifies a bodily experience. This is a type of learning that emphasizes non-words. Hawai'ian children often receive life learning from elders who themselves learned the importance of non-word relaying communication. As Carbaugh (2007) describes a learning cycle from nature, "Nature speaks, we listen, we somehow learn, we struggle to put what we have learned into words, but we are forever frustrated by the process, thus we return to nature and forever enjoy the spiral—..." (p. 68).

Hawai'ian language is also one language of Mother Earth. The ancient chant of Hawai'ian creation, *The Kumulipo*, honors magnificent interconnections with humans and the Earth (Beckwith, 1951, p. xiv). The highly polysemic language contains layers of meanings and symbols connecting subtle and metaphoric worlds. Listening is an indispensable part of such a connection, and it is not a separate or independent tool to connect with nature. Hawai'ians believed in "the power of '*olelo*,' words or language" (Kanahele, 1986, p. 123). Humans could connect with nature by observing, feeling, tasting, and touching, with all the senses that allow humans to be a part of the natural world. Language plays a critical role in calling out and evoking the spirit of nature.

Listening to the voice of nature in daily life starts with ocean waves, air movements, bird and wind songs, and breeze tones through palm leaves. For modern humans it constantly requires what Sewall (1995) calls "the radical awareness of interdependence" (p. 203). For a Westerner, listening to natural phenomena works only metaphorically. Roszak et al. (1995) laments: "In our culture, listening for the voices of the Earth as if the nonhuman world felt, heard, spoke would seem the essence of madness to most people" (p. 7), but for Hawai'ians this is a perfectly reasonable and natural thing to do. Even modern Hawai'ians are capable of communication between man and nature, and nature and supernatural worlds (Heighton, 1971, p. 21). Listen to an experienced surfer from Oahu who reminds us of this:

> Have you ever heard our ocean breathe? We should be learning the heartbeat of our ocean—listen to the breath. She is a living entity and if you listen you can hear the ocean sigh and take a deep breath. People are so unaware of the minute changes that occur in nature that the animal world is in tune with.
> (Pohaku Stone, Surfer, O'ahu, quoted in Provenzano, 2007, p. 58)

Breath, or *ha* as Hawai'ians call it, in his remark is not limited to human breath but is one of nature and the ocean. It is the source of life and all the natural beings surrounding humans. Hawai'ian's breath is expressed in the sound of *ha*, and the actual meaning is "breath of life" or "divine breath." Listening should start with this fundamental activity of life survival. Hawai'ian scholar Lewis (1983) explains that Hawai'i was originally called "Ha-va-iki" (p. 13). *Iki* stands for "small" and *va* for "place of time." When the ancient Polynesian navigators found the islands of Hawai'i, they named it as "the small place of Divine Breath," or "small place for the preservation of the knowledge of Breath of Life" (Lewis, 1983, p. 13). The word *Hawai'i* still is proudly pronounced as /ha-va-i'i/ by locals, emphasizing the main element of the word, which is the breath of life, *ha*.

Ha is the first thing a Hawai'ian recognizes and listens for in another. Hawai'ians greet in a traditional way called *hono ihu*, translated as "nose kissing" or "kiss nose," which is also seen among other South Pacific islanders, Maoris, and Eskimos. When two Hawai'ians greet by *hono ihu*, they bring their noses and foreheads closer to each other to breathe in the same air while gently placing hands each other's shoulders. They literally listen to each other's *ha*. It is "a close face-to-face sharing of the breath of life" (Berney, 2000, p. 26). It also symbolizes oneness by mingling the same air, an act of receiving and giving the divine breath. When an elderly *kahuna* or shaman dies, he passes on the power and knowledge that he had by symbolically "breathing it to his protégé" (Berney, 2000, p. 26).

The famous Hawai'ian word *aloha* is said to have at least 40–50 different meanings (Pukui & Elbert, 1986, p. 21), if not hundreds. *Aloha,* just like *Hawai'i,* contains the fundamental element of *ha* combined with the word *alo*, which means "be with" or "go with." When a Hawai'ian says *aloha* to you, it actually means to "be with the divine breath." "Divine breath", *ha*, is used in other words such as *mahalo* or thank you; the original meaning was "may the breath of divine stay with you." A story goes that when Hawai'ians first saw Westerners greeting or praying, they were surprised to see that the Westerners did not seem to use genuine *ha,* and decided to call them "people without the divine breath, or *haole*" (Lewis, 1983, p. 14).

Hawai'ians do not distinguish "listening" as a separate entity from seeing, touching, smelling, or feeling to recognize the world and natural environment. Listening is a part of learning, understanding, and in-taking. Listening is a gateway to life, survival, and connection to higher being. In her book of Hawai'ian epistemology, Meyer (2003) describes hearing or listening as a part of mystical mediums that expand notions of empiricism (p. 104). She analyzes Hawai'ian perceptions through their unique language expressions and shows us how listening is directly connected to the divinity as breathing is, as in *ha*.

> *i'ke* (sight): to see and to know
> *a'o* (taste): to teach, to learn
> *ha* (smell): to breathe
> *ho'olono* (hear): to hear, to pay attention, invoke *Lono*
> *'a'apo*(touch): to grasp with the mind
>
> (Meyer, 2003, p. 104)

Hawai'ian language in fact has several words for hearing and listening: *lohe, lohena, ho'olohe,* and *ho'olono*. The first three words indicate hearing sounds and listening to each other at human level. The word *ho'olono* is particularly relevant in our discussion as it means listening and paying attention, as well as invoking a spirit and deity (Meyer, 2003, p. 163). The word indicates that Hawai'ians have been well aware of listening to the voice of divine and nature. *Lono,* who is being called upon by human listening, is one of the major deities in Hawai'ian mythology. *Lono* governs a wide range of jurisdictions such as rain, fertility, agriculture, music/sound, as well as intellect of learning, war, and peace. Most importantly, *Lono,* to whom the action of listening occurs, is a god linked with the Earth (Berney, 2000, p. 153). Legend tells that the god *Lono* descended from heaven through a rainbow to help humans, and met his wife, *Laka,* a mortal woman, who later became a symbol of hula. When a Hawai'ian listens, it is the act of asking the Earth for guidance and wisdom. *Lono* is still worshiped and celebrated in the traditional sports and harvest event called *Makahiki,* the Hawai'ian version of the Oktoberfest or the Greek Olympiad.

Meyer (2003) postulates that listening is a part of Hawai'ian epistemology, knowledge acquisition through body, and metaphorical experience. She says:

> Listening... becomes something that is lifted beyond the mundane. To pay attention, to really listen (*ho'olono*) is to invoke a spirit, a deity. *Listening then becomes a spiritual act* [emphasis in original].... It is intimately tied to other and to how we invoke our own genealogy to learn what is most critical. Listening well is found in the act of focus, and focusing is part of what culture helps to define.
>
> (Meyer, 2003, p. 163)

It is interesting to view all these ancient notions about communication once again from the ecopsychological perspective. Harvard zoologist Wilson (Wilson & Kellert, 1993) coined the term "biophilia" (p. 4), which he defined as "the innately emotional affiliation of human beings to other living organisms" (p. 31). In other words, people's behavior, mental state, emotional sanity, and/or insanity are bonded to the surroundings as much as to the state of the environment. When we view this notion from a Hawai'ian perspective, it is the act of listening and feeling and sensing nature that connects humans with nature and gods. Listening is an indispensable bridge between the two seemingly separate worlds. The Earth is deeply interconnected with the psychological and spiritual states of humans as much as the surrounding Earth influences humans.

Ho'oponopono: Listening to the inner voice

Another central notion of Hawai'ian life and communication is a concept of *ohana* or family. *Ohana* in Hawai'i is something much larger than the immediate or extended family in the mainland American culture. Hawai'ian *ohana* includes those whom one is connected to, either by blood, marriage, adoption (*hanai*), or living in proximity, such as people studying in the same school and working in the same organization. It seems to me that *ohana* generously includes a larger community, and even the natural environment, that nurtures one's physical and spiritual life. One native Hawai'ian friend told me that he considers all his ancestors a part of *ohana* too. It means every being, both alive and dead, is a member of *ohana* for Hawai'ians.

Hawai'ian life, however, is not all rosy and conflict-free. Just like any other human society, even *ohana* could be bothered by disruptions, disputes, and discordance. What does one do when one's voice is interrupted, misunderstood, or ignored? How does the wisdom of *ohana* in paradise keep harmony and balance? The answer is that harmony is retained by means of *ho'oponopono*, a traditional mediation system. The fundamental belief of this special mediation and healing practice is that disharmony occurs when the basic homeostasis is disrupted. Hawai'ians think that human imbalance is mainly due to the imbalance of homeostasis between people and environment, which is the larger *ohana*. In the traditional method, various ritualistic steps are taken to restore the original harmony. Let me briefly summarize the method.

Ho'oponopono has been used for hundreds of years to restore balance between family members and a larger community. All the members of the *ohana*, children and elders included, are expected to attend to the ritual once it is called. Even in this modern age, family members are requested to fly across the Pacific Ocean or drive long distances to participate in this ritual. In order to restore balance and harmony, people are required to be truly present and hear out each other's inner voice. In *ho'oponopono*, listening is an essential and indispensable element in seeking each other's forgiveness. Some readers might be reminded of the Quaker religious ritual, the Listening Circle, which may not be dissimilar to the Hawai'ian tradition. *Ho'oponopono* is regarded in Hawai'i as being at least as important for a family as a wedding, birth, or funeral might be for mainland culture.

In a typical *ho'oponopono*, family elders or *kupuna* and/or shaman *kahuna* conduct or facilitate the ritual. The first important step is for the participants to chant together a sacred prayer *E Ho Mai* (see beginning of chapter) while in a standing circle and holding hands. The chant is expected to call forth the ancient wisdom and knowledge that their ancestors also learned and chanted through their generations of experiences. The harmonized chant evokes inner *pono* that opens heart space to bring forgiveness. A distinguished scholar of Hawai'ian tradition, Mary Pukui, explains: "the word *pono* in *ho'oponopono* stands for responsibility and righteousness" (Pukui, Haertig, & Lee, 1972, p. 60). The word *ho'oponopono* literally means "to make things right," which means returning things to harmony and maintaining good relationships among family.

Following the *E Ho Mai* chant seeking wisdom from the family's guardians and ancestors, a problem or issue is presented and identified. Berney (2000, p.115) illustrates these steps:

1. Identifying the problem,
2. Stating the transgression and discussion,
3. Recognizing the negative entanglement,
4. Sharing of feelings and confession,
5. Releasing the problem,
6. Cutting off the problem,
7. Summarizing and reaffirming bonds and closing prayer.

Some *ho'oponopono* sessions could take hours or even days to hear out each other's entangled feelings and emotions, but Hawai'ians see it as a necessary and indispensable way of listening to each other's voice. Days are spent to truly find the core of the issue and, once resolved, it returns to nature and is not mentioned again.

I was fortunate enough to witness one session of an *ohana*'s *ho'oponopono* on the Big Island of Hawai'i. A group of a dozen family members were gathering in a small isle called Coconut Island (or *Moku Ola*) near downtown Hilo. The ancient Hawai'ians considered the island a sacred place, where a wounded enemy soldier was spared if he had managed to escape to the isle. The healing power on the tiny isle adjacent to the Liliuokalani Garden is still enjoyed by the modern park strollers in the Hilo area. The members of the dispute stood in the circle and prayed to each other for forgiveness. It was a solemn and sacred scene, yet children were playing around the family under the swaying palm trees as ocean waves were surrounding the *ohana*. The *ho'oponopono* ritual serves to restore not only the self and family equanimity, but also environmental balance and harmony; as a facilitator says, "the small body serves the larger body" (Bezilla, 2012).

Conclusion

I first became interested in Hawai'ian ways of communication, language, rituals, and spirituality many decades ago. At that time, in the early 1980s, people were concerned about the slow disappearance of Hawai'ian language and traditional culture. It was estimated only 2000 native Hawai'ian speakers remained and they were more than 60 years old (Shook, 1985, p. 3). With the strong and persistent campaigns of Christian missionaries since the mid-17th century, Hawai'ian language had become almost obsolete except in hula or ritual chants. Children on the island became unable to comprehend their grandparents' tongue. The word *aloha*, for instance, became a mere greeting expression. Hollywood and Elvis's *Blue Hawaii* (Wallis, 1961) superficially represented cultural tradition. Hawai'i appeared to have become a tourist attraction, another Disneyland in the South Pacific.

I recently moved back to Hawai'i to teach at a small campus of the state university. Soon I was pleasantly surprised to discover the active movement of cultural restoration. Youth with Hawai'ian heritage are encouraged to take their language courses, receive scholarships, and participate in various cultural and historical programs. The movement is currently well under way thanks to diligent and patient work by concerned Hawai'ian and supporting people. Cheerful exchanges in authentic Hawai'ian language by both Hawai'ians and non-Hawai'ians are daily scenery on this small campus. Language restoration seems to bring back pride and cultural identity to Hawai'ian locals along with realization of the traditional ways of connecting with the natural environment.

Abundant evidence suggests that the Hawai'ian tradition of interconnected life is the key to the renowned Hawai'ian world. It is a way, possibly the only way, to restore wholesome communicating with our Gaia, the wounded Earth. As the "Gaia Hypothesis" posits, because "not only people are connected and shaped by the environment, but we are 'bonded emotionally to the Earth' " (Roszak et al., 1995, p. 13), we need to heal and find harmony while working on the environment.

Berney (2000) tells us what we could do:

> The earth is speaking constantly. Spend a few minutes every day simply listening to all the sounds around you. Hear the machine sounds and separate them out. Then, listen to the voices of the earth: the cry of a bird, the chattering of a squirrel, the whispering of the wind, the sound of rain. Throughout your day, listen for natural sounds. After a while, you will discover subtle patterns that emerge from them. Become sensitive to nature's signs. Let the cries of seabirds foretell rain. Let crickets tell you the temperature. (p. 190)

We all can be humble students of Hawai'i and the ancient wisdom of other native cultures and join the ecopsychological movement. The Earth does matter because our being, our communication, our survival deeply depends on the environment. It is not only the politicians' or environmentalists' job. It is our job, yours and mine, and it is our survival.

Appendix A: Vocabulary—Hawai'ian glossary

a'o	instruction, light, new shoot
a'apo q	to learn quickly, catch the meaning quickly
alo	front, present
aloha	love, affection and 40 other meanings listed
ha	breath
hanai	foster, adopted child
haole	White person, Caucasian, foreigner
ho'oponopono	to correct, to put it right, adjust
ho'olohe	hear, obey, mind, listen, feel
ho'olono	listen
honi ihu	kiss nose, touch nose, smell, scent

huna	hidden secret
ike	to see
kahuna	priest, sorcerer, magician, doctor
kupuna	elders
Kamulipo (The)	Hawai'ian mythology telling a creation of human world
Lono	Diety, one of the major gods brought from Kahiki
mahalo	Thanks, gratitude
Makahiki	Ancient festival beginning about middle of October and lasting about 4 months with sports and religious activities. Taboo war. Now replaced by Aloha Week.
mana	supernatural or divine power, miraculous power
moku	island
ohana	family, kin, relative
ola	life, wellness, well-beings
olelo	language, speech, words, quotation; speak
pono	to be right
pule	prayer, magic spell, incantation, blessing

References

Abram, D. (1995). *The Spell of the Sensuous: Perception and Language in a More-Than-Human World*. New York, NY: Vintage Books.

Beckwith, M. (1951). *The Kumulip: A Hawai'ian Creation Chant*. Honolulu, HI: University of Hawai'i Press.

Berney, C. (2000). *Fundamentals of Hawai'ian Mysticism*. Berkeley, CA: Crossing Press.

Bezilla, A. (2012, September 27). Author's personal communication with Allysyn "Auntie Aloha" Bezilla.

Carbaugh, D. (1999). "Just Listen": "Listening" and landscape among the Blackfeet. *Western Journal of Communication, 63(3)*, 250–270.

Carbaugh, D. (2007). Response to Cox, quoting "the Environment": Touchstones on Earth. *Environmental Communication, 1(1)*, 64–73.

Dudley, M. (1990). *Man, Gods, and Mature*. Waipahu, HI: Na Kane O Ka Malo Press.

Heighton, R. (1971). *Hawai'ian Supernatural and Natural Strategies for Goal Attainment*. Doctoral dissertation, University of Hawai'i at Manoa.

Kanahele, G. (1986). *Ku Kanaka Stand Tall: A Search for Hawaiian Value*. Honolulu, HI: University of Hawaii Press.

Lewis, A. (1983).*Clearing Your Lifepath through* Kahuna *Wisdom*. Las Vegas, NV: Homana.

Maxcy, D. (1994). Meaning in nature: Rhetoric, phenomenology, and the question of environmental value. *Philosophy and Rhetoric, 27(4)*, 330–346.

Meyer, M. (2003). *Ho'oulu: Our Time of Becoming*. Honolulu, HI: Native Books.

Provenzano, R. (2007). *Kai: Ocean Wisdom from Hawai'i*. Honolulu, HI: Watermark.
Pukui, M., & Elbert, S. (1986). *Hawaiian Dictionary: Hawaiian to English, English to Hawaiian*. Honolulu, HI: University of Hawaii.
Pukui, M., Haertig, E., & Lee, C. (1972). *Nana I Ke Kumu (Look to the source)*, Vol. I. Honolulu, HI: Hui Hanai, Queen Lili'uokaklani Children's Center.
Rogers, R. A. (1998). Overcoming the objectification of nature in constitutive theories: Toward a transhuman, materialist theory of communication. *Western Journal of Communication, 62(3)*, 244–272.
Roszak, T., Gomes, M., & Kanner, Al. (Eds.) (1995). *Ecopsychology: Restoring the Earth, Healing the Mind*. San Francisco, CA: Sierra Club Books.
Salvador, M., & Clarke, T. (2011). The Weyekin Principle: Toward an embodied critical rhetoric. *Environmental Communication, 5(3)*, 243–260.
Sewall, L. (1995). The skill of ecological perception. In T. Roszak, M. Gomes, & A. Kanner, (Eds.), *Ecopsychology: Restoring the Earth, Healing the Mind* (pp. 201–215). San Francisco, CA: Sierra Club Books.
Shook, V. (1985). *Ho'oponopono: Contemporary Uses of Hawai'ian Problem-solving Process*. Honolulu, HI: University of Hawai'i Press.
Wallis, H. [producer & director] (1961). *Blue Hawaii* [motion picture]. USA: Paramount Picture.
Wilson, E., & Kellert, S. (Eds.)(1993). *The Biophilia Hypothesis*. Washington, DC: Island Press.

12
Response Essay: Environmental Voices Including Dialogue with Nature, within and beyond Language

Donal Carbaugh

Introduction

In *The Practice of the Wild*, Gray Snyder encourages readers to exercise "grace," a kind of reflective practice that regularly acknowledges what we have been given, its role in sustaining us and in helping us move further along our way. I must begin, then, by giving thanks to Jennifer Peeples and Stephen Depoe for bringing these essays together, to the authors here for their thoughtful words, to those whose words inspired theirs, and to the world we humbly share, which inspires us, and within which we strive to understand our places, as it variously makes itself known to us. This is what sustains us; and for that, I give thanks.

In the following I would like to begin by putting forth a few words of my own about our world. To do so, I will begin by making some familiar observations. At multiple levels, our nature is full of diversity. From deserts to cool northern climates, we live in diverse places, each with its own diverse resources. From central Africa to northern Canada, from the outback of Australia to Lake Baikal in Russia, our geographic sites are so diverse. Similarly diverse are the peoples who inhabit such places. Each has developed its own ways of dwelling in their homelands, in their scenes, and to be sure, those ways are not the same, from the markets of Saigon to Islamabad. When looking carefully within we see diversity among a people; when looking across we see diversity between peoples in those places. Adding to the complexity is the means people use to make collective sense of where they are and of whom and of what they are. Using the Urdu language to make sense of Pakistani places is one thing; using Arabic is another (with my English here being yet another). To make sense of the Mackenzie River in the Yukon through Urdu is quite another. The Finnish language is wonderfully molded to its peoples and places, where English is to some degree serviceable but certainly not revealing of meaning in the same ways. Diversity is a concept that helps remind us of complex systems at multiple levels at once ecological, geographic, communal, and linguistic.

As we add the range of regions to the packs of people to the lots of languages we reach a kind of exponential consideration of diversity to complexity that is in its ways both marvelous and perplexing. The marvel is at the variety or bounty before us; the perplexities become pronounced through our familiar, often inadequate ways of dealing with the differences from place to place, people to people, language to language, and in doing so we not only try to understand what is different to us, but we do so differently through our customary ways of dealing with difference. In the process, complexity often yields to perplexity.

Three current trends accentuate and magnify this dynamic before us. The first, communication technology, has made connecting places, peoples, and languages easier, even as understanding across them has been lacking. Connections are quicker than ever, as are the confusing results of those connections. In the Boston marathon bombing of April 2013, images and words spread quickly and for prolonged periods of time people knew not what they were seeing or hearing. Were those gunshots we heard in the images before us or not; are those people perpetrators or spectators? Judgments were made quickly; understanding was in short supply. Social media, especially tweeting, tended, or trended, to amplify that dynamic.

A second trend adds to our contemporary scenes and that is human migrations. Movement of people has accelerated, also making contacts among peoples more frequent, living out-of-place more typical, being unsettled even a new norm, with interlingual contact familiar to many. With this rapid movement of information and people have come not only deep difficulties in dealing with diversity, but also the vivid demonstration that we are all to some degree interdependent. Our places, peoples, and languages are often tied inextricably together.

These trends have demonstrated to us and our fellow peoples the keen degree to which we depend upon each other, our peoples, our worlds, and our world. In the face of such diversity in places, peoples, and languages, and in the wake of such trends, there is an attendant need for knowledge of not only our own but other places, not only ourselves but others, not only through our language but a stance that is actively attentive to others', to our nature, and to how all of this connects us to our world and its diverse resources. For these profoundly important reasons, we must expand what we mean as we listen to people and languages, to places, by considering the range of possible voices involved, including environmental voices; we would do well to cultivate our abilities to listen carefully to the range, to dialogue with the range in view, to enhance our acting and critical capacities, giving each full consideration.

This is not an easy task, of course, but it is an immediate condition or context for our words here in this volume and for the nicely reflective essays in this particular unit by William Homestead, David Tschida, and Yukari Kunisue. I cannot for a minute pretend to some grand integrating commentary about their rich remarks. What I will do is write a few of my own, many echoing theirs, and I will do so around several guiding themes which I bring together through the following thesis: given the current conditions among places, peoples, and languages, we need an enlarged, enhanced yet situated view of our interconnections which we can call dialogue; this view must endeavor to bring together in an integrative way typical dualistic or binary concerns that are too often kept apart, such as people–place, nature–culture, and human–spirit; this integrative view can be built productively through communication if it is considered an activity that lies within yet beyond speaking, and includes listening to nature; and listening as such must be actively attentive to a diversity of voices or a variety of potentially powerful communicative agents including nature, as among all things, Thoreau's turtle, Olson's lakes, and the incomparable Hawai'ian Islands. Thoreau's turtle, Olson's lakes, and the incomparable Hawai'ian Islands. Let us begin.

Dialogue as a guiding theme

As a part-time resident for decades of both Finland and Blackfeet country—the latter a Native American place in northern Montana, USA—I have come to admire practices in each which tie people productively to their places. Some readers will undoubtedly know that Finland is a Scandinavian nation which ranks at or near the top of various eco-friendly indexes, no matter how these are measured. Accompanying this lifestyle is a penetrating ingredient in the Finnish culturescape which was put succinctly into English by one Finnish commentator, Markku Koivusalo. As he says, a Finnish stance involves "the quiet observation of the world expressing itself" (1999, p. 49). Imagine that: a culturally valued stance is declared which involves "quietude" and attentiveness to "the world" as a complexly expressive partner. The active stance here can be engaged in a deeply thoughtful silence, identified in one way in Finnish as *mietiskellä*. To this complex world, through this periodic Finnish stance, we find we should listen thoughtfully and deeply as an engaged partner. What an idea this is.

This is an idea not born simply of heady philosophical abstraction, but of everyday practice. A summer seasonal if not weekly ritual of many Finns is a trip to the cottage or forest. Part of the trip is not simply recreational if it is that at all, but reflective of the world in which one lives. As a Finnish friend of mine, after much sitting in silence, put it to me as we enjoyed a sauna at his cottage, "being here brings the nature into my senses, the sight and sounds of the birds, and the water. Here I can enjoy that in a good way." This practice, and this sort of practice even if in a park in Helsinki's or Turku's harbors, helps connect people productively to their places, attentive to its qualities, voices, dynamics. While it is a mistake to leave the impression that all Finns value and regularly engage in such a practice, it is the case that many do at some time, and that it is active often at a cottage or park, or even as friends enjoy moments of silence together. This is a means of connecting people, of attending to places, and of learning about each other's and one's own connections to it all as "the world expresses itself." And this latter is a key part: treating "the world" as an expressive partner.

Regarding this point, Markku Koivuslao wondered: "what can Finnish stillness offer?" He invites his reader to question with him the value in this sort of contemplative practice. Just what good can it do in this world today, as active as it is with globalization, immigration, fast-paced technology, instant and perpetual contact being possible with, it seems, nearly everyone? His response:

> Perhaps it can remind us of the limits of the words used at the [common] table and the ephemerality of human culture. It can emphasize the significance of wordless nature outside the realm of table conversations and the need to care for common things and offerings. One need not see silence simply as an empty passivity, but it can also be seen as a reserve of strength. Stillness may bring to the table practical design and matter-of-factness.
>
> (1999, p. 53).

And in this stillness, the world expresses itself, yet this is not a fanciful or only pantheistic holy silence—although in Finnish the phrase *pyhä hiljaisuus* can embrace that quality too—but it is also a means to "practical design and matter-of-factness," two deeply Finnish qualities—the latter codified in Finnish as an *asiallinen*, a person who is direct in approach and focused frankly on the matter at hand (Wilkins, 2005). Such a style as valorized in Finland and Finnish can open new and direct ways of getting things done with an air of productive simplicity; and in getting things done by listening in stillness to "the world." If we believe the eco-indicators, among other important practices, there is some possible large scale value in that (Carbaugh, Nurmikari & Berry, 2006).

There is another practice elsewhere which shares some features with this Finnish one. This is evident at times among traditional Blackfeet people on the northern Great Plains of the United States. As the Blackfeet origin story is told by the ancestral figure, Napi, people are advised, when having difficulty, to listen to the animals such as eagle, buffalo, or bear. These creatures can carry deep meanings that may speak constructively to one's troubling circumstances, but one has to be focused in a good way to get the corrective message. Such a message might come in one's sleep, in a ceremony, or in a reflective moment, but one has to be ready for it when it comes, or it will be missed. And being ready means one listens also to agents in the world which are nonhuman. The world does indeed express itself. Hearing it in this way is a way of cultivating a power in one type of traditional listening (Carbaugh, 2005).

This sort of ancient practice puts speaking itself in space which alternates with an expressively active "silence." The Pawnee and Otoe poet Anna Lee Walters has written about this as follows (1992, p. 13):

> Silence and speech at the water's edge
>   alternated here
> Remember that we need both
>   we are told
>
> The totem voices took turns speaking
>   the bears roared
>   the eagles screeched
>   the pigeons cooed
> Until all the clans had spoken
> Silence followed

> *Silence*
> Then all the totem beings spoke at once. Yes,
>   through their speech and voices
>   and through the ensuing silence
>   the people, the clans knew they lived.
> This is the power of language
>   but often it is not realized until silence prevails
> *Silence*
> *Remember both, we are told*

The Finnish practice demonstrates the value of this sort of practice in a highly industrialized, superbly literate, technologically saturated nation; the Blackfeet practice demonstrates the value of the practice as rooted in this one case in a contemporarily situated ancient wisdom. The chapters in this volume by Homestead, Tschida, and Kunisue give further witness to the nature and meaning of kindred types of this practice as seminal in Henry David Thoreau's writings, Sigurd Olson's ideas, and among native Hawai'ian people(s). Each suggests deep value in vigilant attention to our places and peoples as these make themselves known to us; each provides access to knowledge we may not capture by other means; each is valorized as a social good, both intrinsically valuable as an action and in the outcome of a deeper understanding. If we only learn to so listen, and as we do, we engage productively with our places and peoples; this being a sort of dialogic stance which holds promise for social and ecological betterment and thus enhances our ability to act well into the future. We seem to be on to something important!

William Homestead writes tellingly about Thoreau and his relationship with his woods and world. At one point, Homestead writes about "dialogic wisdom" as a kind of stance-in-the-world through which we are opened to an understanding or knowledge which through other means, may be missed. Bringing Martin Buber's thoughts into his discussion, Homestead reminds us of the difference in treating our addressee as a non-organic, objectivized "it," rather than a valued and sacred "thou." The latter can even be applied, he argues along with Buber, to "diverse species as articulating subjects." We find robust grounds for this dialogic wisdom as we become kindred spirits with ancestors, Thoreau and Buber, who are, in Homestead's words,

> advocates for prerational, sensuous listening to the voices of nature; scientific—naturalist inquiry into the voices of relationship and

reciprocity, and transrational experience of the *Voice* of nature, apprehending a spiritual unity among material diversity. All are necessary if we are to fully embody Buber's claim that living means being addressed; all are necessary for an ethical responsiveness to ecocrisis.
(this volume, p. 184)

There is much to think about in such a statement. And I want to think with Homestead about this. First, when discussing a Thoreauvian-style dialogue, Homestead distinguishes an "animistic orientation" from a "transcendental spirit." In his words, "such participatory consciousness reflects an animistic orientation, in which everything that is, is alive, rather than a transcendent Spirit within all things." As I read these words, I wonder about the value in such a distinction between the animistic and the transcendental. On the one hand, things are deemed in some sense alive (of this world); on the other, they are understood to be spirited (of another?). Places and peoples vary in the degree to which such qualities are said to be part of their world. Which are active and to what degree do these seem valuable for us to know? An assignment of the one rather than the other gives voice to some more than others, some features of place more than others, amplifying this and muting that. I think Homestead may be the first to say a wise dialogic stance seeks to bring both dynamics into view and does not require an absolute resolution, with this sort of reflection itself being born of such a stance.

The issue, however, can be a thorny one. In prior and present times, religious wars have been fought over such matters. Today with the intimate interdependence of places and peoples, we must learn as our worlds express themselves, and this dialogic stance may help offer a way of doing so. What to do next? One cannot say in any absolute sense. But what to do here, in this place, among these people, considering its situated knowledge, now we are at a situated point Koivusalo has called local "practical design," in its place, here-and-now practice rather than everywhere-and-always dogma (Carbaugh & Cerulli, 2013).

Second, Homestead writes about Bateson's ideas of mind and nature and at times the mind–nature dualism. A point to recall, initially, is that such a distinction is part of a Western mindset, and that this is amplified in a Germanic tradition, as expressed here in English. This geo-linguistic orientation assumes a kind of rationality for its point, and for the existence of the dualism. Thoreau himself was keenly influenced by this sort of Germanic thought (Richardson, 1986, pp. 20–31). Other points about the matter brought into consideration from elsewhere are cast differently.

Homestead nicely captures how Thoreau built according to such a view:

> Thoreau advocates for prerational, sensuous listening to the voices of nature; scientific—naturalist inquiry into the voices of relationship and reciprocity, and transrational experience of the *Voice* of nature, apprehending a spiritual unity among material diversity... All are necessary if we are to fully embody Buber's claim that living means being addressed; all are necessary for an ethical responsiveness to ecocrisis.
>
> (this volume, p. 184)

Should we separate the natured world from the human world, the concrete from the idea?

In Thoreau's sense, dialogue involves deeply serious and sensuous listening, pre-thought and pre-rational experience, caught yet connected within one's human, natural, spirited environs. Listening and learning in this or these ways helps one embrace a kind of existential unity among diversity, and vice versa, better equipped to respond to the very conditions of one's place, of our places. Through this, the pre- and transrational are rational, the dualism of mind–nature may be unified, while at the same time the unity may be diversified. The dialogic attitude is important to emphasize: not one way for all, but in each place, its own and best ways.

Third, what do we say when we say, with Thoreau, "spirit aware of itself"? Those smitten with John Muir, as am I, will recall a favored tagline in so many of his "wilderness essays," of nature as "varied expression of God's love." In so seeing, and so feeling, we encounter "spirit aware of itself." Or in Thoreau's killing of a turtle, we somehow grapple, as Thoreau did, with apparent inadequacies in so encountering our variably expressed worlds. If it is the case, as Homestead writes, that "everything speaks of spirit," I want to add immediately that it does so diversely by place, people, and voice. I worry the language of "unity" coupled with "spirit" leads easily down singular roads of theological dogma or the like, and thus to closing various doors into the houses of peoples and places we might usefully enter.

Finally, as a student of communication, I found myself reading Homestead and wondering this: what model of communication do we use when we say "all things speak" or we claim to "listen in dialogue with nature"? Homestead quotes Thoreau on the futility of the matter: "Perhaps the facts most astounding and most real are never communicated by man to man" (1982, p. 463). Or as Ralph Waldo Emerson put it earlier: "We know more from nature that we can at will communicate." Within one certain logic of communication, limited to human language alone, with its attendant muting of the world, our efforts are indeed deeply constrained. Tutored in other expressive means such as those that introduce this chapter, and the others discussed here, we can enrich and variably enliven our understanding of what communication involves. One of the humbling lessons students of modern communication must learn is the depth of ancient wisdom about what communication is and can be, as captured in the poetry of Anna Lee Walters above, and others who so practice. There is indeed ancient wisdom alive in the world today if we can hear it.

With this point in mind and with the materials discussed above, it is important to note that Thoreau wrote his essays as he was composing his "Indian books" (see Keith, 2007). Indeed his chief biographer called this "Thoreau's lifelong fascination" (Richardson, 1986, p. 219). In them, he kept detailed records of the ways indigenous peoples in North America and around the world knew their worlds and spoke of them. His well known writings take on renewed indigenously inflected meanings once the reader knows of this, Thoreau's "fascination." Deep threads of these practical ways are alive in some of our modern views of communication, as they are active in Thoreau's writings, and as they are in the previous and following remarks, but they are oft muted in many, many others.

The movement to a listening-based model of communication challenges us to contextualize language use as one means of expression beside various others including our nonverbal ways (e.g., Scollo, 2004). As Homestead nicely reminds us,

> Thoreau's bias is integrative, informed by experience beyond words and thoughts, remaining open to mystery and free from totalization, and thus it reflects a better bias. And anthropomorphism, when abused rather than reflecting an empathy that reveals relational insights, should be questioned via critical rationality informed by other modes of knowing. Sympathy with intelligence does not lead to anthropomorphism, but to an awake and widened awareness.
> (this volume, p. 200)

Through a variety in means of communication, we can come to embrace such wisdom of diversity among places, peoples, and voices, and the wisdom spans not only peoples, but the variety of worlds, the nature in which we live.

On dialogic listening

We have been reflecting upon potential value not only in listening, but in listening as a means of accessing knowledge about nature, worlds, places, and peoples. David Tschida writes about this process as "dialogic listening" and—in his review of works by Lipari and Tompkins—as activating four qualities: an ethical obligation to be open to difference or an Other; an attitude or value in mutually beneficial care, or compassion; a willingness to go beyond one's habits, prejudices, or typical modes of action; and being open to view the world as another sees or feels it. We might think of this as cultivating an empathic ability to listen as the other listens, and thus understanding better one's stance in the process. Both are needed for dialogue, and we see eventually that transformation is possible through the process.

Tschida nicely reflects primarily upon the works of Sigurd Olson. Olson lived among the great northern lakes of Minnesota which he paddled by canoe in summer and skated in winter. His deep love of these places equipped Olson to become President of both the Wilderness Society and the National Parks Association. He worked similarly to help draft the federal Wilderness Act. His way with words attracted many people to his beloved lake region while opening his readers to experiencing or listening to these places anew. What is this path Olson has opened for so many, this "dialogic listening"?

Tschida captures several qualities of this form as it is active in Olson's writings. It is a multi-sensory activity; one listens to the wilderness through all of the senses. In doing so, one may be able to capture a sense of awe or wonder in nature's presence (as at Olson's Listening Point); one may hear and see things wholly anew. It is noteworthy that this sort of dialogic listening carries with it a humbling quality; one realizes deeply that one is merely one small part of it all. At the same time, there is the potential realization in listening that the wonder of nature somehow benefits us all. As nature lovers and writers variously say, including Homestead and Kunisue, there is a realization of "sympathy with intelligence" or a connecting with "earth wisdom." At this listening point, Olson emphasizes one should be willing to be surprised, that is, the mystery of nature may make itself known in surprising ways. Coming full circle, we may reconnect to that of which we are already part, discovering in the process, as Olson puts it, "the wonder and peace" which is already there but which we somehow missed or if we caught it before, forgot. As Tschida puts it, as we listen dialogically, we may enter a process of "being in communion and longing."

This process in the abstract may sound rather mystical to many readers. For that reason, I find it productive to dwell on it, to bring it home literally, to render it as a radically emplaced activity. After all, we listen somewhere and our typical places are our everyday ones. This connects with Tschida, I think, when he writes about an "extensionist view." This view suggests listening with and to our wilderness scenes in ways its voices can be understood. I think this process is most typically done in the concrete scenes of our typical activities rather than in the abstract. I have been asked often if I could give an example of this and when I do, it always starts: "When I am in Montana at this site..." or "When by this stream behind my house..." and so on. This activity is by its very nature situated, emplaced. Another kind of response would be to mention, or read as I have my students do, John Muir's fantastic essay about the water ouzel or dipper. After reading it, ask yourself what type of listening Muir did in order to write that essay. Now that is radically emplaced dialogic listening, multi-sensory, full of awe, beautifully humble, a sympathy with intelligence broadly conceived, eye-opening, wondering and learning par excellence. Such "communion and longing" in Tschida's sense, the "awe and wonder" of Olson, is a naturally emplaced practice.

Whether this ability connects with a "heavenly sphere" or world beyond, I cannot be sure. I do not doubt, however, that it connects with a place and its wonder. Embracing the diversity in such places, knowing what they may offer, and also knowing when they are violated, including those who live there, is valuable in itself, an emplaced listening and enhanced learning about what is among us, here. Cultivating how to listen dialogically to nature, in nature, with nature and all it holds, as Olson and Tschida do, is an invaluable action, and through it we can be open to difference, be compassionate, and not only learn things anew, but learn anew how to so learn.

On indigenous ways: Listening to nature

Those who study indigenous peoples' ways have often noticed parts of traditional practices which deeply link people to places. A most celebrated work along these lines is Keith Basso's (1996) *Wisdom Sits in Places*. Through detailed ethnographic examinations of communicative devices such as place-names and stories, Basso demonstrates how western Apache people not only identify features of their landscapes but also use them as moral guidance for everyday living. For example, among western Apache people of Cibecue, Arizona, when a person is having a difficult time or is in some way troubled, an interlocutor can mention a place, such as "by the bend in the river," with this place-naming act being followed by a prolonged period of silence. During this period of silence, there is intensely active communication. To those familiar with this practice, the place-name brings with it a story which is presumably known to the troubled person and the interlocutor. The troubled person is invited to think about that story. In it is dramatic action which reveals a moral, such as "recall that one should act with caution and that others care deeply for you." The person who is troubled, then, can take comfort in the message of the tale, all of this transpiring through a contemplative silence. This morality tale then is actively invoked through the place-name with the upshot of the communication being advice and comfort for a troubled person.

Communication such as this is readily available among western Apache people as a part of everyday social interaction. Various place-names are known and with them various stories. Each story provides guidance for living. But western Apache people do not need others to name places to remind them of the stories and the morals they teach. They also have their landscape before them, standing ready as reminders of proper living, as places which hold stories and thereby express proper moral premises in living a life well and in living well with others and properly in their places. In this sense, as Basso deeply reveals, the land "stalks" those who live there through these morally infused ways; it serves as a constant reminder through a complex expressive system of morality tales. Traditional indigenous ways such as these link people deeply to places, to proper living, to each other.

Yukari Kunisue recounts another indigenous set of practices which operate similarly in tying people deeply to each other and their places. She explores a native Hawai'ian voice as a way of being with nature, as a way of indeed hearing nature. We are introduced by Kunisue to some features of modern life in Hawai'i which manage "to sustain the spirit of *Ha*, wisdom carrying *Mana* (life force), and *Huna* (the secret wisdom of life)." One such indigenous form of communication which cultivates these Hawai'ian premises is "talking story," a narrative account of lineage and place which radically situates the storyteller in a precise physical, moral, and communal environment.

Kunisue's description notes that several everyday occasions invite this and similar forms of cultural communication. In her words,

> people in their uniquely modest way tell me their stories (Hawaiians call it "talk stories"), share mythologies, and ancient legends. I hear chanting and *Pule* (praying) on the lava beach on an early Sunday

morning. I view authentic *Hula* dances, both on stage at the internationally acclaimed Merry Monarch Festival, or simply practicing in the park. They look deeply awed and proud, carrying the heritage and being nature themselves. I observe Hawai'ian *Kupuna* or elders greet each other by *Honi Ihu* (nose kissing). I learned an ancient mediation method, *Ho'oponopono*, conducted by skillful elders. I see all these rituals are a Hawai'ian way of listening to the voice of their natural environment.

(this volume, p. 229)

The scene is set through a complex system of expressive practices which connect people to each other and to their natural places.

From outside, gestures such as "nose-kissing" and "hula dancing" may seem removed from nature's places. Yet when placed as part of a system of traditional Hawaiian practices, each playing a role in the creation of the other, and each cultivating a way of living together-in-place, one can start understanding the deeper links and cultural premises. Through such practices, here on the islands, people keep a mythos of stories, an ethos of feeling about its places and its people alive. Through these ways in these places a people are living a traditional wisdom.

Kunisue's account is avowed as an ecopsychological one. She brings a view of a people about their place into view. She recounts ethnographic details such as the belief that fish are potential participants in communication and thus one should be careful around them; the fish might hear the intent of fishermen and not offer themselves as food.

As Kunisue notes, there is an indigenous view of communication that is active here. It connects people in its own ways to places in which they have dwelled for centuries. This includes engaging in dialogue with the animals who live there with them, the organic matter around them, patterns of climate, and geography. Through this view, as Kunisue notes, wisdom speaks from all around; and all of this is indeed spirited. Kunisue reminds us that "Hawai'ians worship natural phenomenon in their daily life by offering respect and awe to hundreds of deities. Communication takes form in wider perceptions and sensations (this volume, p. 230)." We are reminded that communication of this indigenous kind is not only through words or images. There is more to it than that. The "more" is difficult, then, to capture in words. And herein, in this traditional way, lies what Kunisue calls a "paradox": "as soon as humans speak about nature it [that form of communication] 'alienates humans from the material-physical reality of nonhuman nature'" (this volume, p. 240, citing Maxcy, 1994, p. 330).

At this point, we reach the limit of written words. All we can do is point with them, beyond them, to "nature" (here intended beyond the word itself) and confess that there is more there than we can say, but much there that we can learn to know. Perhaps the point is more ironic that paradoxical—after all these words are being devoted to it—but either way the point stands: indigenous communication such as this takes us beyond language about nature, to nonlinguistic ways of listening with and to nature. Through these ways we know more than our words can tell. Privy to "the more," we are positioned to more wisely speak the words we use (Carbaugh & Boromisza-Habashi, 2011).

This dynamic in indigenous communication, as with elements in Thoreau's and Olson's thoughts, introduces elements of ineffability and mystery which are often seen as counterpoints to prominent voices of Western science and logic (Berry, 2005). There is an underlying or overlaying or immanent (the more metaphors the better) condition for such traditional means of communication: the world speaks in a spirited way and we should listen to that even if we cannot ably put—in an exact way—what it says into our own words. I think there is a deep, productive lesson in humility in this which we should heed. As Wendell Berry has put it in the title of a recent work, it is "the way of ignorance," of recognizing that there is much we do not know, and we would do well to proceed accordingly, cautiously, humbly, rather than with an attitude of hell-bent certainty. This sort of listening offers insight and knowledge which we most likely cannot gain in any other way.

A key feature in Kunisue's report, and in similar others, is the introduction of metaphorical language and what indeed is believed to be literal, and what metaphorical. It is one matter to say the "earth speaks" as a metaphor; it is quite another to say it is literally true. Kunisue makes the point explicit, following Roszak, Gomes, and Kanner (1995, p. 7): "In our culture, listening for the voices of the Earth as if the nonhuman world felt, heard, spoke would seem the essence of madness to most people." In traditional Hawai'ian practices, however, this sort of communication is apparently deeply active. Such acts keep people, nature, and spirit together, in their places. The Hawai'ian premise of interconnection, *Ohana*, brings spiritual life, physical place, and one's ancestors together into one's community. Listening to that is important. Learning what to say based upon that is a necessary condition for good living. A key objective is to maintain good relations among all that there is.

Among Blackfeet people in northern Montana there is a traditional story, perhaps even the first story cast as it is from the "time before time." It starts by placing its listeners in a troubled place, as being overcome by life's circumstances. In need of assistance, one is encouraged to cry aloud for help, perhaps through a prayer or in a dream. If successful, one may receive the favor of a reply, an offering; it is typically from an animal such as an eagle, buffalo, or bear. The story tells its listener that you must listen to the animal, "obey it," and "be guided by it." This can offer you, dear listener, what you need to get by. Indeed, the final line says, "that was how the first people got along in the world." For many people today, and many before us, the world indeed does speak.

Have we forgotten (how) to listen?

In dialogue with nature

My remarks here are added to those in this volume in an effort to invite a sort of emplaced dialogic stance with regard to nature, its places and peoples. Brought into view are voices we typically silence, whether in other languages or from nature itself. As we join this conversation, we can be enlivened by it. With Homestead and Thoreau, we can situate ourselves in places to listen and learn from them; with Tschida and Olson, we may discover awe and wonder in nature's ways, and in the ways others learn of awe and wonder; with Kunisue and Basso, we find a humble, ancient wisdom, guided by listening to and learning from nature more than speaking one's mind about it.

Again, given so many of our current conditions in places, among peoples, and between languages, we will benefit by enlarging and enhancing our views of nature and our worlds. This may help us bring together in a more integrative way concerns that are too often kept apart, which are typically deemed polemical, dualistic, or binary: people versus places, nature versus culture, and human versus spiritual matters. A dialogic view can be built productively through communication if it is considered an activity that lies not only within but beyond speaking, and this includes listening to nature. As such, listening to nature-in-place, with others, can be actively attentive to a diversity of voices or a variety of potentially powerful expressive agents. Perhaps this will open us to a sort of "practical design" in Koivusalo's nice phrase if we cultivate our ways of hearing the world expressing itself, and guide our actions accordingly.

References

Basso, K. (1996). *Wisdom Sits in Places: Landscape and Language among the Western Apache*. Albuquerque, NM: University of New Mexico Press.

Berry, W. (2005). *The Way of Ignorance*. Berkeley: Counterpoint.

Carbaugh, D. (2005). *Cultures in Conversation*. London and New York: Erlbaum Publishers.

Carbaugh, D., Berry, M., & Nurmikari-Berry, M. (2006). Coding personhood through cultural terms and practices: Silence and quietude as a Finnish "natural way of being." *Journal of Language and Social Psychology, 25*, 203–220.

Carbaugh, D., & Boromisza-Habashi, D. (2011). Discourse beyond language: Cultural rhetoric, revelatory insight, and nature. In Christian Meyer and Felix Girke (Eds.), *The Interplay of Rhetoric and Culture* (pp. 101–118). Oxford and New York: Berghahn Book Studies in Rhetoric and Culture III.

Carbaugh, D., & Cerulli, T. (2013). Cultural discourses of dwelling: Investigating environmental communication as a place-based practice. *Environmental Communication: The Journal of Nature and Culture, 7*, 4–23.

Keith, B. (2007). Thoreau's "Indian books" and Bradley Dean's "broken task": A personal remembrance. *Thoreau Society Bulletin, 260*, 1–3.

Koivusalo, M. (1999). Still life: The aesthetics of Finnish silence. In *Finland: The Northern Experience, New Europe and the Next Millennium* (pp. 48–53). Helsinki: Tammi Publishers.

Maxcy, D. (1994). Meaning in nature: Rhetoric, phenomenology, and the question of environmental value. *Philosophy and Rhetoric, 27(4)*, 330–346.

Richardson, R. D. (1986). *Henry Thoreau: A Life of the Mind*. Berkeley: The University of California Press.

Roszak, T., Gomes, M., & Kanner, Al. (Eds.) (1995). *Ecopsychology: Restoring the Earth, Healing the Mind*. San Francisco, CA: Sierra Club Books.

Scollo Sawyer, M. (2004). Nonverbal ways of communicating with nature: A cross-case study. In S. L. Senecah (Ed.), *The Environmental Communication Yearbook, Volume 1* (pp. 227–249). Mahwah, NJ: Lawrence Erlbaum.

Snyder, G. (1990). *The Practice of the Wild*. San Francisco, CA: North Point Press.

Thoreau, H. D. (1982). *The Portable Thoreau*. Ed. C. Bode New York: Penguin.

Walters, A. L. (1992). *Talking Indian: Reflections on Survival and Writing*. New York: Firebrand Books.

Wilkins, R. (2005). The optimal form: Inadequacies and excessiveness within the "asiallinen" [matter-fo-fact] nonverbal style in public and civic settings in Finland. *Journal of Communication, 55*, 383–401.

13
Coda: Food, Future, Zombies
Eric King Watts

I wish the reader to know something up front that I think is vital for an appreciation of my involvement in this project: I had never conceived of voice as anything but a human phenomenological problem. And although two-thirds of the essays making up this collection focus on human-based agencies, the volume enjoys the company of three elegantly written meditations on what I am calling a "spirit-voice." When I was approached to pen an epilogue for the work, I admit that I was concerned that I would have to encounter a notion of voice that did not comport with my own and I would, thus, need to confront my own understanding of the happening of voice—an event that I had thought of as occurring through the human–human relation. I imagined this encounter to be a fearsome problem, but it turns out to be a problem of the very best kind. In this final turn in the volume, I will first relocate voice to the site of my latest obsession—zombies. By doing so I hope to begin a reconceptualization of voice that might be possible in the human–zombie relation; this revision allows me to start the confrontation that I allude to above. Second, the human–zombie relation implicates primal issues discussed in this volume—food crises and the deterioration of the planet. Finally, I want to suggest how the timely essays gathered here provide interpretive tools and interventionist practices necessary for forestalling a zombie apocalypse looming on the horizon due to rapacious eating of each other and our shared Earth.

In "Mourning for the Earth: A faith-based response to climate change," Katharine M. Preston (2013) wonders about a deafening silence that envelops speech about the fate of the planet. It is not that we do not talk about climate change; rather we neglect our feelings regarding our planet and, therefore, we refuse to talk about the relations of "loss, grief, and climate change" (p. 59). Every writer in this collection knows that some folks patently deny climate change itself. I had a conversation recently with a gentleman that I consistently run across at the YMCA, where the topic of violent storms was being discussed. His faith-based response to disasters like Hurricane Sandy was that God's retribution was at hand. God was angry and I had nothing to say about it. This friend was silent on the subject of whether he himself was sad over such catastrophes or whether God's wrath was engineered through climate change. And so I was struck by Preston's insistence that we consider that "some part of God" (2013, p. 59) erodes along with the environment. Like William Homestead, Yukari Kunisue, and David Tschida in this volume, Preston is committed to animating a concept of voice that is sacred: "Have we ignored our emotional and spiritual connections to the planet? Could the noise swirling around climate change—science, politics, media blitzes, as well as the weather disasters themselves—drown out the voice of a loss so profound that it rests unnamed in our souls?" (2013, p. 58). The authors collected here are clearly seeking interventions on two dimensions alluded to by Preston—the noise and the voice. But first, let us talk about voice and zombies.

The intersection among voice, zombies, and the environment is so obvious to me now that I'm embarrassed that it didn't occur to me long ago. In a recent work (Watts, 2010), I took up the challenge of animating zombie voices as a way to appreciate a mode of politics that increasingly (and once again) treats people as "things." The backdrop for this dehumanizing activity is a kind of wasteland. You are probably familiar with the narrative conventions of the contemporary zombie: some kind of infection or disease is at the heart of people succumbing to sickness. Somehow they rise from the dead, existing in a nether space/time that is predominantly understood as undead or living dead. Zombies shamble about towns and countryside, drawn to the smell and sound of the living; apparently their complicated motives and desires have been reduced to a single drive to eat. The zombie embodies world-shattering contagion and a biomechanical delivery system for bringing about the apocalypse. Zombies are also speech impaired. But they do make sounds—moans and gurgles and chomping noises that are reportedly too terrifying to endure (Brooks, 2003). Zombies are in a state of slow decay and degeneration not of their making. One way to understand the zombie is they are forms of falling flesh that we must destroy and make quiet so as to suspend the "call" they utter. Rather than smother this sound, I have become increasingly invested in reanimating zombie voices.

Voice is the sound of affect. Voice emanates from the openings that cannot be fully closed; from the ruptures in sign systems, from the breaks in our imaginaries, from the cracks in history. It registers a powerful, some would say passionate cluster of feelings triggered by life finding a way to announce itself. In my earlier conceptualizations (Watts, 2001; Watts, 2012), this announcement required an endowment by hearers or publics daring enough to acknowledge the affective and ethical dimensions of speech-in-the-making. I say daring because there is no guarantee that voice will be welcomed or pleasing; just as often, voice shocks and incites violence and hatred. And so, voice signifies a connection that is fraught with tremendous uncertainty; it marks the space/time of a dangerous dialogue. It is not difficult to imagine the conditions where people can perceive voice as seditious, heretical, or sacrilegious. Voice tends to upset the status quo. With these notations in mind, let us consider what zombies can show us about environmentalism and the works in this collection.

To even think of reanimating zombie voices one must make two adjustments. First, one must suppress the urge to shoot zombies in the head. Second, one must come to some sort of understanding of the undead. The former adjustment requires impulse control while the latter change entails a brief history lesson regarding the plight of the Afro-Caribbean zombie slave. North American and European attraction for the zombie can be traced to imperial projects. The key point is that Haitian folk practices have been partially involved in zombification rituals for hundreds of years and these rituals seized the imaginaries of European colonialists because they seemed to satisfy an entire range of beliefs, fears, and desires regarding the savage (Davis, 1988; Sheller, 2003). Importantly, the zombie slave was not exactly dead brought back to life; the zombie slave was under a spell and some tales relate the undead capacity for the zombie slave to revolt and throw off the shackles. What accounts for this potential in the zombie slave? It is the same force that always lurks within those who depend on masking their motives and feelings for survival, looking for the right moment to flee or fight. Perhaps the moan of the zombie signals this force and we should risk its acknowledgment, endowing voice.

Voice does not make a specific statement; it announces inchoate feelings. What we dread about zombie voices is the intimate contact with the thing we are supposed to destroy or dominate. Zombie voices, therefore, might function as a call of conscience regarding what we imagine them to represent—disease, posthuman dreams or nightmares, the unrecognized desire to make others work for little or nothing.

The zombie moan might also express "a loss so profound that it rests unnamed..." In this way, the zombie figures ecological degeneration writ large. If I can argue that zombie voices alert us to the disposability of humanity that is becoming more common, why not revisit the human-nature relation at the heart of "spirit-voice"? Zombie voice is the voice of the not-quite-human; or perhaps it is the sounds of profound prejudices making this "not-quite-human" perspective seem reasonable.

Bracketing my prejudices regarding the essentially humanistic character of voice may better attune me to what William Homestead asserts, using Thoreau, as the "voice of nature." Homestead rightly contends that I do not extend voice to nature or spirit, but sees no reason for this truncation, offering instead different ways of thinking about dialogism. His conceptualization seems to hinge on linking pre-rational, rational, and transrational experiences "transcending the dualism between mind and nature." We have long been involved in monological interventions in the ecology of the planet rather than entering humbly into dialogue with it. Yukari Kunisue, in "Listening to the Natural World," takes a similar tack, but utilizes a different methodology. Beginning with ethnographic descriptions of Hawaiian listening practices, Kunisue wants to reawaken in us an "ancient understanding" that we have (or used to have) access to, and that we can utilize to cultivate "transhuman dialogue" with our environmental surroundings. Kunisue makes this key claim: although nature is always expressing its aliveness, each listener makes their own meaning. This understanding is consistent with my notion of voice—it enunciates a relation that can be accepted or denied, and we are "free" to state the terms. It is precisely our refusal or incapacity to accept this invitation always already being offered, or accepting it in the most exploitative and selfish terms, that these essays engage. David Tschida, in "The Ethics of Listening in the Wilderness Writings of Sigurd F. Olson," seeks to explore why we tend to refuse nature's solicitation. Through a recuperation of this nature advocate's writing, we not only get an interesting snapshot of his life's pursuits, but we receive a cogent answer to the question of why we should care about attending to the call of nature. An appreciation of voice, Tschida argues, draws one toward the radical contingency of one's own existence; one's situatedness clarifies one's connectedness with nature, one's ensnarement with the human–nature relation. I have been involved thus far with working through a revision of voice made palpable by the human–zombie relation and recontextualized as human–nature relations. These encounters with voice in this collection are correlated with confrontations with the "noise" that often drowns it out.

Voice and Environmental Communication can be understood as having three relatively distinct threads. The first thematic that I addressed was "spirit-voice." The second and third dimensions I call "advocacy-voice" and "critique-voice." There are four stimulating essays that make up the "advocacy-voice" category. Two of these works, Benjamin Garner's "Vote with Your Fork," and Leah Sprain's "Voices of Organic Consumption," capture the urgency of the politics of food and environmental action. This thread in the volume still echoes my interest in the zombie apocalypse since such an undead zone sponsors critical concerns over food production, distribution, and safety. Indeed, the zombie figures food frights that harken famine (Newbury, 2012). Garner mines the site of a specific farmers' market using interviews with patrons to draw out motives for supporting slow and local food. Garner conceives of agency in terms of buying power and views the commercial support of these markets that practice sustainable ecology as a form of "political voice." The interviews that he conducted bring to light two recurring attitudes among market-goers: (1) a fear of environmental decay linked to mammoth agribusiness and (2) a "disdain for those responsible." Sprain's essay shares a commonplace with Garner's; each posits consumptive acts as political acts. Also, each writer is interested in what patrons of "organic" consumption might say about their own buying habits. Sprain, however, surveys posts to Usenet.com regarding what counts as organic and why such definitions matter to marketing and food politics. Her essay offers a cloudier picture of everyday food advocacy as practiced by patrons because there are few viable standards shared by all consumers.

The second two essays I'm grouping under "advocacy-voice," Casey Schmitt's "Invoking the Ecological Indian," and Jessica Prody and Brandon Inabinet's "Sustainable Advocacy," take up specific case studies testing both the theoretical conceptions and practices of advocacy. Schmitt focuses on the mobilization of a racial trope called the "Ecological Indian." As a trope of the noble savage, the Ecological Indian is closer to nature than the modern subject and thus has been pressed into the service of environmental campaigns. Schmitt rightly assesses this (mis)use of the Ecological Indian as over-determining the way Americans envision actual Indians and their cultures. He wonders how voice is even possible under conditions where this trope dominates the public imaginary. The activist and orator Winona LaDuke serves as a case for grappling with how this trope can be reshaped and re-sounded so as to allow authentic voice to come through, especially regarding a wide range of life-threatening problems like housing and healthcare. Schmitt posits the case of the trope as a manifestation of a problem that a revised theory of voice might remedy. Prody and Inabinet move from the theoretical to the case. In "Sustainable Advocacy," they explore how theories of environmental sustainability impact models of advocacy. In particular, these scholars are concerned with how intergenerational audiences might be constituted and mobilized through public address. The case here is a speech by a Canadian youngster, Severn Cullis-Suzuki, at the 1992 United Nation's Earth Summit. The rhetorical theories of Perelman and Olbrechts-Tyteca allow these writers to imagine the little girl's youth as a kind of temporal bridge between the "logical" long-term project of environmentalism and "practical" short-term considerations. Voice here is a way to attend to those needs and desires of audiences not yet in existence.

The final two essays making up this volume occupy the category of "critique-voice" because they each entail sustained analyses of high-profile agents that seek to thwart environmental advocacy campaigns. Peter K. Bsumek and co-authors, in "Corporate Ventriloquism," offer a smart analysis of how the coal industry and associated interests use front groups like wooden puppets; these advocates "throw" lobbying voices through the front groups to make them seem autonomous and unified in their support for coal. To the well established corporate strategies that manufacture consent for increasingly unpopular ideas like "astroturfing" and "greenwashing," the authors propose we add "corporate ventriloquism." I think they are on to something here, but would suggest that ventriloquism is not "voice" per se, but voicelessness. This slight but significant conceptual modification of voice could also help an otherwise solid discussion of Michael Crichton's public attacks on the North American environmental movement. In "Defending the Fort," Patrick Belanger seeks to disclose the appeal that the famous science fiction writer marshalled while undermining green projects. Belanger explores two of Crichton's voices—patronizing and populist—as instrumental to his rhetorical agency. I want to suggest that voice might be able to do more in these two nice essays (and the volume as a whole) if it is not reduced to speaker agency or style, as it occasionally seems to be. For example, the ventriloquism that Bsumek et al. identify could use the authors' intervention in the form of an endowment of the voice that is muted. By acknowledging the affects and ethics of the speech of the front groups, we might be able to discern more robustly the interests that are not heard, the persons and communities choking on the dust of coal and made to stand up for the polluter or else. The authors are certainly marching in that direction; I only want to enable that trajectory.

Voice and Environmental Communication is a challenging and provocative volume; it brings together diverse perspectives and methodologies. It does not shy away from asking us (asking me) to change ways of thinking about and relating to the world. This work helped me to be less pessimistic about our planet's future and the resources it offers us. My interest in zombies has been fueled by the dysfunctional political culture around which our future selves orbit seemingly helplessly. The strength of this collection lies not in its pieces, its rigorous theses ranging from spirit to advocacy to critique; rather, its potency more precisely can be appreciated as a response to that unnamed loss that we must acknowledge in order to recover it.

References

Brooks, M. (2003). *Zombie Survival Guide: Complete Protection from the Living Dead*. New York: Three Rivers Press.
Davis, W. (1988). *Passage of Darkness: The Ethnobiology of the Haitian Zombie*. Chapel Hill, NC: University of North Carolina Press.
Newbury, M. (2012). Fast zombie/ slow zombie: Food writing, horror movies, and the agribusiness apocalypse. *American Literary History, 24,* 87–114.
Preston, K. M. (2013, November–December). Mourning for the Earth: A faith-based response to climate change. *Utne Reader, 180,* 58–60.
Sheller, M. (2003). *Consuming the Caribbean: From Arawaks to Zombies*. New York: Routledge.
Watts, E. (2001). "Voice" and "voicelessness" in rhetorical studies. *Quarterly Journal of Speech, 87(2),* 179–196.
Watts, E. (2012). *Hearing the Hurt: Rhetoric, Aesthetics and Politics of the New Negro Movement*. Tuscaloosa: University of Alabama.
Watts, E. K. (2010). What is this "post-" in post-racial, post-feminist... (fill in the blank)? *Journal of Communication Inquiry, 34,* 214–222.

Index

academic engagement, 177
activism, 79, 120–1, 148, 151, 167
  as performance, 165
agriculture
  community supported agriculture, 153
  environmental impact, 158–65
  industrialized/corporate, 150–1, 158, 167
  organic/alternative, 127–9, 137, 166–7
  sustainable, 79–80
animal mistreatment, 161
anti-intellectualism, 49
Appalachia, 21–2, 26–35, 39
appropriation, 22, 24–7
  aggressive mimicry, 25–6
  astroturf campaigns, 25–7, 28–9, 36, 38, 114
  lateral appropriation, 25
argument spheres
  personal, 44, 45, 63, 65, 108
  public, 44, 45, 50, 61, 63, 65, 108
  technical, 44, 45, 49–50, 56, 61, 63, 64, 65, 108
Asen, Robert, 67, 70, 86
attitude
  diplomatic, 96–8, 106
  logical, 96–8, 106
  practical, 96–8, 106
  rhetorical, 96, 107
audience, 28–32, 35–8
  intergenerational, 89–93, 95–107
  particular, 78, 89–90, 92–5, 98, 100–2, 105–6
  universal, 12, 89, 93–6, 100, 105–7, 118

Backes, David, 205, 215, 225
Bakhtin, Mikhail, 4–5, 15
Bateson, Gregory, 184, 186, 202
Berry, Thomas, 10–11, 183–4, 202
Berry, Wendell, 187, 202, 254, 255
biophilia, 221–222, 224, 235
  defined, 235
Blankenship, Don, 31–2
Brummett, Barry, 208, 209, 225
Buber, Martin, 184–5, 193, 202, 246–8
Burke, Kenneth, 88, 107, 132, 144, 209, 222, 225

Callicott, J.Baird, 222, 224, 225
Carbaugh, Donal, 10, 13–14, 15, 72, 86, 100, 106, 108, 189, 199, 201–2, 206–7, 225, 229–30, 232, 239, 245, 247, 254, 256
Carson, Rachel, 4, 6, 17, 63, 212–13, 225
citizenship, 32, 40, 170, 175–6
Clair, Robin, 142–3, 144
Clarke, Tracylee, 10, 17, 67, 72, 87, 188, 203, 230, 240
climate change, 39, 40, 46–7, 53–4, 57, 63, 64, 65, 77, 78, 86, 91, 97, 107, 108, 109, 110, 112–13, 115, 118, 121, 122, 196, 203, 257–8, 263
  Intergovernmental Panel on Climate Change (IPCC), 47, 64, 65
  manufactured controversy, 113–14
  science, 47, 49–50, 59–60
  skepticism, 44–65, 112–13
Coakley, Carolyn Gwynn, 205–6, 227
coal
  Appalachian Voices, 29, 39
  coal industry, 12, 21–2, 24–40, 112, 262
  Faces of Coal (FACES), 22, 27–30, 41, 42
  Friends of Coal (FOC), 27, 41, 42

264

Massey Energy Company, 31
West Virginia Coal Association
(WVCA), 27, 43
colonization/colonialism/imperialism,
59, 75, 77–8, 85, 259
decolonization, 70, 86
common sense, as argument, 23, 44,
50, 51, 55
consciousness-raising, 102–4, 106–7
consumerism, 30, 37, 116, 117,
127, 129
consumption, 6, 13
of coal, 29, 30, 36
conspicuous consumption appeal,
138, 140
as lack of voice, 140–1, 142
of organic food, 127–4
as political action, 127–31, 135–8,
141, 143–4
tasteful consumption, 130,
138–40
as voice, 128, 130–2, 133, 141–3
corporate ventriloquism, 21–40
defined, 22
Couldry, Nick, 1–6, 14, 22–4, 26, 30,
36, 37–8, 48, 93–4, 104–5, 111,
118, 142–4, 150, 165–6, 171
Cox, J. Robert, 1, 8, 15, 25–6, 41, 115,
120, 122, 206, 226
credibility, 44, 45, 48–9, 60–2,
73, 101
expertise, 45, 49–50, 55, 59
Crichton, Michael, 12, 44–65,
112–13, 262
Cullis-Suzuki, Severn, 90–1, 97–106,
108, 118, 262
cultural performances, 152, 165–6

deep ecology, 50–1, 186–7
DeLuca, Kevin M., 7, 16, 223–4, 226
democracy, 7–8, 39–40, 89, 101, 136
Athenian, 88
dialogue
dialogic attitude/stance, 185–6, 187,
189, 246–8
dialogic ethics, 221–3
dialogic event, 4
dialogic listening, 13, 205–7,
209–12, 218–21, 250–1

dialogic theory, 184, 188
dialogic wisdom, 184–6, 190, 194,
197, 246
dialogue with nature, 189–90, 192,
202, 255
transhuman dialogue, 188, 228, 260

Earth Summit Address (1992), 98–9
Ecological Indian, 66–75, 77–8, 80–5,
86, 112, 116, 120, 261
see also Native American
ecopsychology, 13, 186–7, 203,
229–30, 240, 256
Emerson, Ralph Waldo, 190–1, 197,
199, 200, 203, 248–9
environet, 4–5
environmental advocacy, 39,
110–123, 262
environmental communication (as a
field of study), 1–3, 4, 12, 15, 109,
115–16, 118–19, 121, 122, 123,
177, 178, 189, 226, 256
environmental decision making, 8–9,
12, 14, 17, 112, 122
environmentalism, 17, 46–7, 50–6, 61,
62, 64, 96, 110–121, 149, 158–65,
226, 259, 262
"Environmentalism as religion", 44,
46, 50–6, 62, 63
environmental justice, 7, 16, 17, 59,
77–8, 116, 123, 167, 219,
222–5, 226
environmental policy, 5, 8, 23, 45, 61,
88–109
environmental science, 12, 44, 45–7,
53, 56–7, 60–1, 63, 86, 113
environmental social movement, 76,
87, 111, 121, 163, 218–20, 221
environmental campaign, 51,
122, 261
ethnography, 130, 132, 135, 145, 146,
148, 168, 173–5, 228, 251,
253, 260
ethnography of
communication/ethnography of
speaking, 132, 145
ethos, 17, 45, 48, 66, 75, 76, 253
see also credibility

food
    farmers market, 13, 143, 149–50, 153–4, 165–6, 168, 169, 171–2, 174, 175, 176, 177, 261
    food studies, 129–30, 148–50, 175
    local food, 153, 167
    organic, 13, 127–30, 132–5, 137–4, 155–6, 159, 160, 162, 164, 166, 171, 173, 174, 176, 177, 261
    Pollan, Michael, 135, 146, 149, 169, 173, 179
    see also agriculture
Finland/Finnish people, 242, 243–5, 256
Fisher, Walter R., 48–9, 61–2, 63
front groups, 22, 26, 40, 262
    see also appropriation

generation, *see* intergenerational *under* audience
global warming, 46–7, 57–8, 64, 65, 110, 113, 122, 123, 201
    denial, 95
    see also climate change
Goffman, Erving, 138, 145, 168
Goodnight, G. Thomas, 44, 46, 49, 50, 62, 63, 92, 104, 108
grace, 240
grassroots organizing, 26–9, 36, 38
greenwashing, 25–6
    see also appropriation

Hawaiian language
    Ha (divine breath), 13, 228, 231–5, 238, 252
    Ho'olono (to hear, to pay attention), 234, 238
    Ho'oponopono (Hawaiian mediation ritual), 13, 228, 229, 235–7, 238, 240, 253
    Huna (the secret wisdom of life), 228, 231, 239, 252
    Lono (Hawai'i an deity), 234, 239
    Mana (life force), 228, 231, 239, 252
    Ohana (family), 235–6, 237, 239, 254
Hirschman, Alfred, 130–2, 136, 140, 141–2, 145

homology, 208–9, 225, 226
human-to-human communication as ideology, 205–8

ideograph, 33, 35, 208–9, 211, 222, 226
ideology, relationship to rhetoric, 36, 134
intergenerational justice, 92, 104–5
    see also audience, intergenerational
internatural communication, 10, 16
    see also dialogue

Kennedy Jr. Robert, 31, 41
Keystone XL Pipeline, 110, 115, 120, 121, 123
Krech III, Shepard, 66, 67, 68–71, 72, 74, 75, 78, 80, 82–83, 85, 86

LaDuke, Winona, 12, 68, 71, 75–82, 85, 86–7, 99, 109, 261
    see also Native American
Leopold, Aldo, 3–4, 9, 16, 71, 212–13, 222, 226
Lipari, Lisbeth, 209–11, 214, 216, 217–18, 219–20, 223, 226, 250
listening
    appreciative, 206
    comprehension, 206
    as constitutive, 211, 218
    critical, 170, 206
    dialogic listening, 13, 205–7, 209–12, 218–21, 250–1
    discriminative, 206
    ethical/moral obligation, 207–8, 209, 212, 213, 219–21, 225, 250
    human-to-human, 205–6, 207, 211, 216, 218–19, 223, 225
    human-to-nature, 219, 223–4, 260
    moral sensitivity, 209–10, 227
    to nature, 184–90, 201
    as a socially constructed concept, 206–7, 211
    speaking, bias toward, 210–211
    therapeutic, 206
    typology of listening, 206
Listening Point, 213–14, 221
    Ely, Minnesota, 212–14, 216, 250

McGee, Michael Calvin, 208, 209, 226
media, 16, 22, 27, 32, 43, 46, 49, 57, 64, 65, 68–9, 104, 171, 226, 242, 258
Meyer, Manulani, 233–5, 239
Milstein, Tema, 207, 224–5, 226
Mitra, Ananda, 4, 16
Mother Earth (Gaia), 16, 81, 232, 238
Muir, John, 1, 2, 3, 16, 71, 207, 226, 248, 251
Myerson, George, 4–5, 16

Nadasdy, Paul, 67, 72–4, 77, 81, 82, 85, 87
narrative/narration, 6, 12, 13, 15, 28, 44, 48–50, 51–2, 53, 56, 61–2, 63, 83, 86, 89, 94–5, 114, 134–5, 144, 150, 155–8, 162–4, 172, 211, 217, 221–2, 224, 252, 258
  counter-narratives, 190
  dominant narratives, 183, 185, 189, 190, 201
  Fisher, Walter R., 48–9, 61–2, 63
  narrative style, 48–9, 62
Nash, Roderick Frazier, 221, 226
National Environmental Policy Act (NEPA), 8
Native Americans
  Anishinaabe, 68, 75–81, 87
  Blackfeet, 10, 15, 72, 86, 108, 189, 202, 207, 222, 223, 225, 229–30, 239, 243, 245–6, 254
  Ecological Indian, 12, 66–85, 86, 112, 116–17, 120, 261
  noble savage trope, 66, 69, 70, 261
  Ojibwe, 78
  stereotype, 66–8, 70, 74, 75, 82–3, 116–17
  Traditional Environmental Knowledge (TEK), 71–5, 81–2, 83, 87
  Weyekin principle, 10, 17, 72, 87, 188, 203, 230, 240
  White Earth, 75–6, 78
nature
  as agent, 188, 205–6, 243–5
  constitutive approaches to, 188–9, 203

dialogue with nature, 189–92, 202, 255
dualism, 185, 186, 188, 193, 196–9, 201, 247–8, 260, see also nature: monologic attitude
listening to nature, 99–100, 184–90, 228–41
monologic attitude/knowledge, 184–6, 188, 189, 197, 202
nature as other, 205–8, 212–24, see also dualism
nature's voice, 9–11, 13, 183–202, 221, 231–2, 246–8
speaking for nature, 197–200
voice of nature, 183–205, 260
Weyekin principle, 10, 17, 72, 87, 188, 203, 230, 240
Neoconservative, 22, 26, 28, 32–6, 37, 40
neoliberalism, 15, 17, 22–4, 26, 28–32, 34–40, 41, 42, 63, 108, 111–12, 114, 116–18, 122, 142–3, 144, 146, 165, 168, 176, 178
The New Rhetoric, 93, 96, 109
Nordhaus, Ted, 116–17, 119, 120, 123

Olbrechts-Tyteca, Lucie, 89, 93–4, 96–7, 98, 105, 107, 108, 109, 117–18, 262
Olson, Kathryn M., 218–19, 222, 226
Olson, Sigurd F., 13, 205–27, 246, 250–1, 254, 255, 260
Oravec, Christine, 3–4, 16, 207, 226

pastoral, 33
patronization, 51, 62
Peeples, Jennifer, 5, 7, 16, 26, 42
Perelman, Chaim, 89, 93–4, 96–7, 98, 105, 107, 108, 109, 117–18, 262
performance, 13, 14, 42, 62, 71, 102, 106, 148–9, 151–4, 165–7, 172, 174
Pezzullo, Phaedra, 25, 42, 151, 154
place, 13, 15, 78, 86, 147, 152–3, 165, 169, 171–2, 174, 187, 203, 206, 212–14, 216–17, 221–2, 224, 226, 229–30, 241–55, 256
place-names, 251–2
Plec, Emily, 5, 10, 16, 17

political consumption, 129–30, 131, 135, 137
political engagement, 12, 17, 67–8, 70, 71, 74, 77–8, 107, 116–17, 122, 169
political process, 7–9
Pollan, Michael, 135, 146, 149, 169, 173, 179
  see also food
populism, 48, 51
Preston, Katherine, 257–8, 263
prudence, 50, 54
public argument, 44, 63, 134
public participation, 114–15, 120, 122

reservation, *see* White Earth
rhetoric
  constitutive rhetoric, 13, 134, 138, 144
  rhetorical criticism, 61–2, 63, 130, 132–4, 171, 207
  rhetorical situation, 8, 12, 23, 90, 93, 95, 119, 120, 134
  rhetorical theory, 22, 24, 90–4, 96–7, 104–7, 108, 109, 112–13, 171, 262
Rogers, Richard, 9–10, 17, 70–1, 73, 74, 83–4, 85, 87, 188, 203, 228, 240
Runes of the North, 212, 226
Rydin, Yvonne, 4–5, 16

Salvador, Michael, 10, 17, 72, 87, 188, 203, 230, 240
Schwarze, Steve, 119–20, 123, 137, 146
scientific consensus, 53, 108, 110, 113–14
Scott, Rebecca, 24–5, 32–3, 42
Senecah, Susan, 2, 8–9, 17
Shellenberger, Michael, 116–17, 119, 120, 123
silence, 14, 89, 98, 104, 130, 131, 140–1, 142–3, 144
Silent Spring, 50, 63, 212, 225
The Singing Wilderness, 214, 217–18, 226

social movement, 6, 17, 22, 42, 129, 137, 143–4, 146, 147, 177, 218–20, 223
  see also environmental social movement
social organizing, 5–7, 40, 118
Songs of the North, 216, 226
speech codes, 132–3, 146
spirituality
  spirit-voice, 257, 260–1
  spiritual development/awakening, 190–2, 201
State of Fear, 44–8, 55–60, 62, 63, 65
sustainability, 12, 52, 66, 68, 70, 83, 88–92, 104, 107, 108, 109, 112–13, 115, 117–18, 148, 154, 158, 162–4, 168, 178, 262
  advocacy, 88–107
  science, 177, 178
symbolic violence, 70
sympathy with intelligence, 194–6, 197–8, 200, 249, 250–1

technology as savior, 183–4
Thoreau, Henry David, 13, 183–202, 203, 204, 223, 246–9, 254, 255, 256, 260
Tompkins, Paula S., 209–11, 213–14, 216–18, 220, 223–4, 227, 250
Traditional Environmental Knowledge (TEK), *see* Native Americans
transcendentalism/transcendentalist, 184–5, 186–7, 190–1, 193, 194–6, 198, 199, 200–2, 203, 247
transhuman dialogue, *see* dialogue
transrational, 184, 186, 189–90, 194, 201, 246–8, 260
trinity of voice, *see* voice
Tuan, Yi-Fu, 193, 221, 227

Udall, Stewart L., 69, 205
United Nations Earth Summit, 90, 97, 108, 118, 262

voice
  and credibility, 48–50
  crisis of voice, 11–13, 22–4, 36–40

definition, 1, 130–2
and environmental advocacy, 118–21
Hirschman, Alfred, 130–2, 136, 140, 141–2, 145
and identity, 3–4, 66–85
indecorous voice, 8, 115, 120, 122
of nature, 183–202
neoliberalism and voice, 22–4, 36–9
of nonhuman nature, 9–11, 183–256
and political agency, 150–1
and political process, 7–9
as process, 23–4, 48, 62, 150, *see also* Couldry
as rhetorical force, 60–2
and social organizing, 5–7
as textual/intertextual, 4–5
trinity of voice, 2, 8–9, 16, 17
as value, 23–4, 48, 118, 150, *see also* Couldry
voicelessness, 42, 65, 66, 87, 109, 262

Wasson, Matt, 39, 42
Watts, Eric King, 1, 2, 4, 6, 14, 16, 17, 23, 42, 48, 49, 50, 65, 66–7, 71, 87, 89–90, 94, 99, 109, 111, 123, 131–2, 142, 147, 150, 152, 158, 166, 169, 185, 204, 219–20, 224, 227, 258, 263
website, analysis of, 27–31
Weyekin, *see* Native Americans
Wilderness Days, 205, 214, 216, 226
Wilson, Edward O., 235, 240
Wolvin, Andrew D., 205–6, 227

zombie, 257–63
as conscience, 259–60

Printed and bound in Great Britain by
CPI Group (UK) Ltd, Croydon, CR0 4YY